THE WISDOM OF BIRDS

Three little owls. This species' scientific name, Athene noctua, *comes from the Greek goddess of wisdom, Athene, and its nocturnal habits; its English name comes from John Ray, 1678. Painting by J. Wolf in J. C. and E. Susemihl, 1839–52.*

THE WISDOM OF BIRDS

An Illustrated History of Ornithology

Tim Birkhead

BLOOMSBURY

Published by Bloomsbury USA, New York
Distributed to the trade by Macmillan

All papers used by Bloomsbury USA are natural, recyclable
products made from wood grown in well-managed
forests. The manufacturing processes conform to the
environmental regulations of the country of origin.

LIBRARY OF CONGRESS CATALOGING IN-PUBLICATION DATA HAS BEEN APPLIED FOR

ISBN-10 1-59691-541-2
ISBN-13 978-1-59691-541-1

First U.S. Edition 2008

1 3 5 7 9 10 8 6 4 2

Typeset by Hewer Text UK Ltd., Edinburgh
Printed in Singapore by Tien Wah Press

To Nicholas Davies
for at least thirty years of friendship

CONTENTS

Preface

Somewhat unsteadily, I am standing waist-deep in very cold water in the midst of a vast, open wetland. As part of a team of ornithologists trying to capture one of Europe's rarest and most unusual birds, I am holding a vertical bamboo cane attached to a twelve-metre-long fine net, mounted at the other end on a second bamboo cane and held taut by a colleague. There are three nets in total, and six of us – all in chest-waders – are advancing gingerly in a pincer movement towards a tiny bird singing from a sprig of willow that protrudes from the water. At a predetermined signal we attempt to run towards our target, keeping the nets vertical, but it is like running in treacle and I am aware of how ludicrous we look. Startled by our movements, the little bird makes a bid for freedom and flies between the poles, but as though by magic falls securely into my net. I cannot stop laughing: it seems unbelievable that such a ridiculously inelegant operation could result in the capture of such a special bird.

In the hand, the aquatic warbler is a tiny streak of brown feathers distinguished by a prominently striped black and gold head. As is obvious from the sexual swelling between his legs, this is a male; otherwise he is identical in appearance to the more rarely seen female. We place a single metal ring and a unique combination of plastic rings on his legs, take a drop of blood for DNA analysis, and let him go.

The Biebrza wetland in north-east Poland is magical corner of Europe. Around its edges local people live in Hansel and Gretel cottages and farm the lush water meadows in ways unchanged since the Middle Ages. Unsurprisingly perhaps, this is an area of extraordinarily abundant birdlife. Earlier that day, as I lay in bed, the dawn chorus was almost unbelievable in its volume and diversity: golden orioles, bluethroats, scarlet rosefinches, greenfinches and yellowhammers, as well as swallows, house martins and house sparrows. By night there's a different chorus – less diverse, but no less wonderful: creaking corncrakes, buzzing great snipe and tic-tocking spotted crakes. Throughout the day black storks, cranes, rollers, magpies, harriers and goshawks cruise overhead in this birdwatchers' wonderland.

Out on the marsh, on a metre and a half of brown peaty water, floats a dense blanket of grass and herbs – the aquatic warbler's unique breeding habitat. Invisibly, inside this matrix of vegetation just a few centimetres tall, the females build their nests and, running around like mice, hunt the insects that abound here. For just an hour at the beginning and end of each day the males emerge from this miniature forest to sing from low perches or perform their beautiful song-flights. With such an inaccessible and furtive lifestyle, it is little wonder that this bird has been such a mystery, but now, thanks to the international team of which I am briefly a part, the aquatic warbler's extraordinary biology is slowly being revealed.[1]

Among the majority of birds, males and females work together as a pair to rear offspring – but not the aquatic warbler. Instead there seem to be no bonds whatsoever between the sexes, whose lives consist of apparently random sexual encounters after which the female – in contrast to many small birds – rears the chicks alone. A further difference from most other small birds is the male aquatic warbler's lack of a fixed breeding territory. While the males of many bird species advertise their ownership of a free-hold during the breeding season by singing, aquatic warblers seem instead to drift across the marsh, spending a few days here and few days there with no fixed base, singing simply to announce their availability to the females they desperately seek. DNA fingerprinting reveals that mating is utterly promiscuous and that in many nests every one of the aquatic warbler's brood of up to six chicks has a different father.

Observing birds as elusive as the aquatic warbler in their natural habitats is time-consuming and involves considerable patience, for often there are long hours in which nothing much happens. Whatever species I am studying, I try to use this time to mull over what we know and what it means. Because of its crepuscular activity and unusual biology, the aquatic warbler provides more time than usual and raises many more questions than usual, forcing me to ask where my basic assumptions about birds have come from. The aquatic warbler has made me and the other team members re-evaluate what we previously thought of as 'normal' in birds and to ask where our bird lore originates. Where does our information about the territories, songs, mating patterns and other aspects of the lives of birds come from?

Our knowledge of birds must have started long ago, for in order to hunt them successfully humans had to know something of their behaviour and ecology: where and when they might be at certain times of year; when they reproduced; how they reproduced; whether they nested on the ground or in trees; whether they laid a single egg or many.

The sheer abundance, visibility and diversity of birds means that they have been a source of fascination ever since people began to paint and write. Images of birds decorate the walls of European caves; in Africa the forms of birds were chipped out of slabs of hot, red sandstone; and in Arctic burial chambers the skulls of great auks were placed alongside the dead to accompany them to the next world. The Greeks were inspired and mystified by birds: they wrote poems about them; they employed their body parts, droppings and internal secretions as medicines and used their behaviour to tell the future.

There have been many strange ideas about birds. Some are still familiar, including the belief that certain geese emerged from barnacles attached to logs floating in the sea; or that the pelican sacrificed itself for its young by piercing its own breast and allowing them to feed on the blood; or of swallows overwintering in the mud at the bottom

of ponds. Others are less well known and include the extraordinary notion that a parrot anointed with the secretions from a Brazilian frog would change its plumage from green to red; or that the males of certain types of pigeon can trick females into abandoning their offspring and partner and fly away with them; or that birds can change sex.

Which of these is true? Which is false? How do we distinguish fantasy from fact? When did we start to care? And who was responsible for this change in attitude towards our knowledge?

After the successful capture of our male aquatic warbler the failing light and voracious mosquitoes sent us hurrying back to our primitive lodgings. In an upstairs room of a rented house on the edge of the marsh we prepared a simple picnic of soup and sausages over a small stove and discussed the day's events. We also talked about how much we took for granted and how little we knew about where and who our present knowledge of birds came from. Who was responsible? In an attempt to answer that question, I had during numerous conservations with other ornithologists in similar field situations asked them who they considered the most influential ornithologist of all time.

Our international team was staunchly patriotic. The Germans nominated Erwin Stresemann, the man who in the 1920s professionalised ornithology by bringing together museum studies and field ornithology. The Americans chose one of Stresemann's students, Ernst Mayr, who emigrated to the United States in the 1930s and as a result of his work on the evolution of birds subsequently became known as the 'twentieth-century Darwin'. The British selected another evolutionary biologist, David Lack, who pioneered studies of the ecology and life histories of birds.

When I told my colleagues that, while greatly admiring all of these men, there was someone I rated even more highly, they looked surprised, unable to imagine who it might be. My nomination for the most influential ornithologist went back much further than theirs, to the seventeenth century and to someone who transformed our knowledge of birds from fantasy into fact: John Ray.

A central figure in England's scientific revolution, Ray was more

Previous pages: A misericord from Lavenham Church, Suffolk, dating from the fifteenth century: the pelican piercing its own breast in an act of self-sacrifice to feed its chicks was a popular emblematic image.

than an ornithologist. He was a biologist in the broadest sense of the word: he knew plants, he knew insects, but above all, he knew how to think. He was a philosopher, and it was his mode of thinking about the natural world that changed ornithology. Standing on the threshold of the medieval and the modern worlds, Ray scrutinised and rejected much that had gone before and, looking to the future with extraordinary foresight, anticipated many of the issues that continue to fascinate ornithologists today. Clever and industrious, John Ray was also charming and modest, as his portrait indicates, and you cannot help but like him.

When I told my colleagues that I ranked John Ray higher than Stresemann, Mayr or Lack, to my surprise most of them had not even heard of him. I was disappointed, but also delighted, because it means that the story I'm about to tell in these pages will be new.

Ray's vision of a new natural history, including ornithology, was vast, encompassing everything from straightforward descriptions of what he saw to an entirely new way of thinking about the natural world.

The early 1600s witnessed a massive change in outlook, breaking away from the deeply entrenched ideas of Aristotle, from old wives' tales and from uncertainty. Ray spearheaded a new vision of the natural world, and he did so with a gentle modesty that belied a brilliant mind and wonderful clarity of vision. Inspired by a small group of Cambridge colleagues, Ray overturned the old order in which people saw themselves as guilty sinners living under the watchful eye of an angry and jealous God, offering them something altogether more cheerful: a benign God. Ray's God was responsible for the natural world in all its beauty and in particular the wonderful fit between an animal and its environment – something he called physico-theology (later known as natural theology) and what today we call adaptation. The culmination of his life's work, *The Wisdom of God*, published in 1691, laid out Ray's ideas in brilliant, readable style. In its day, physico-theology was as significant as Darwin's natural selection would be one hundred and fifty years later. Ray's *Wisdom* transformed the way people viewed the natural world. It was a revelation:

7

for the first time there was a scheme that unified natural history and accounted for the way the world was.

After years of studying plants, Ray, with help of a younger colleague, Francis Willughby, became interested in birds. The publication of his encyclopaedia, the *Ornithology of Francis Willughby*, in 1678 (dedicated to Willughby, who had died unexpectedly a few years earlier), set a new standard in the study of birds. Sweeping away the myths and folklore of his predecessors, and focusing on verifiable facts, clear descriptions and careful interpretation, Ray sought certainty of knowledge. Inevitably he made some mistakes; this was after all the very beginnings of what we can call scientific or enlightened ornithology. But by comparison with what had gone before, Ray's *Ornithology* was a breath of much-needed fresh air. The focus was on the 'arrangement' of birds – how they fitted together in God's scheme. The fact that blue tits and great tits were more similar to each other than they were to the northern cardinal (or Virginia nightingale, as it was sometimes called) was obvious. Much less obvious was whether the twite, linnet and redpoll were separate species or simply variations on a theme. Issues like these had tormented and confused would-be naturalists for centuries, but Ray cracked it. By coming up with a definition of what constituted a species, he generated a scheme of naming and arranging that would inspire Linnaeus sixty years later and stay the test of time. Indeed, some considered Ray's system superior to that of Linnaeus. The *Ornithology* was an inspiration and the starting point for an enormous surge of interest in the classification and 'systematics' of birds that continues to this day.

The second thing that Ray did was to provide a conceptual framework for studying birds: his physico-theology was the skeleton around which ornithological knowledge could be built and interpreted. Without a framework, knowledge remains a mere collection of facts, sometimes interesting, but on their own little more than stray feathers. As well as providing a scheme for interpreting facts, physico-theology encouraged those interested in birds to go out and observe them directly. 'Believe what you see, rather than see what you believe' was Ray's message. He inspired others to make their own observations – not easy without the help of binoculars or telescope – and interpret them in an objective way. Ray stood on the threshold of new knowledge: he was the first scientific ornithologist, the first to be concerned about what was true and what wasn't. To sort fact from fantasy, Ray had to reappraise much of what had gone before. Who, for example, among his predecessors had first noticed the frantic autumnal hopping of caged nightingales? Who suggested this was a thwarted southward migration? Who later exploited this hysterical hopping to map the routes and genes underlying migratory behaviour?

Ray's contribution was twofold. His *Ornithology* initiated the study of bird taxonomy, while his *Wisdom* launched the field study of birds – what we would now call the ecology of birds in their natural environment. Either one would have been enough to ensure his enduring reputation. To have done both was extraordinary.

My aim is to use John Ray as the starting point for establishing the headspring of important concepts or ideas in ornithology and to trace their development through time. Scientific ideas are like seeds: depending on who tends them they can flourish and grow into substantial bodies of knowledge, but if they fall on stony ground or find themselves in the wrong hands they can wither away. Some ideas emerge before their time and are ignored, only to be given a second chance later on. Exploring the development of ideas allows us to look back into the rich and fascinating history of ornithology.

The history of ornithology is immense.[2] To reduce it to manageable proportions, I focus on a set of concepts central to the lives of both birds and the enthusiasts who have studied them for the past two thousand years. Concepts are the ideas that hold a discipline – like ornithology – together but at the same time point the way for new research. General concepts that are relevant across a wide range of circumstances or apply to many different species are the most important. My own field of research, the behavioural ecology of birds, has its foundations in the major concept of natural selection and how an individual bird might maximise the number of descendants it leaves for future generations. It was precisely for this reason that I was so fascinated by the aquatic warbler.

While most birds breed as monogamous pairs in apparent marital harmony, appearances are deceptive, for the majority of species – blue tits, red-winged blackbirds and fairy wrens – are no more faithful than humans, and illegitimate offspring are not uncommon. Promiscuity, however, provides the potential for some individuals to leave more descendants than others. One prediction resulting from this is that, as well as seeking extra-pair encounters, males should be keen to avoid being cuckolded. Since the aquatic warbler has taken promiscuity to unprecedented levels, the males of this species should also be particularly concerned about protecting their paternity. And they are, and do so in a rather unexpected way. Because of the species' unusual habitat, females are extremely elusive and can easily evade the sexually overcharged males. This means that when males do encounter a female ready to mate they take full advantage of it and, instead of copulating for a couple of seconds, as most birds do, the male aquatic warbler remains *in copula* with the female for as long as thirty minutes, simultaneously preventing other males from mating at the critical time, but also introducing sufficient sperm to swamp those of any previous (or subsequent) male. This is an evolutionary game of winners and losers, and natural selection favours those individuals that out-compete the others and leave the most descendants.

Today, evolutionary ideas like these guide and direct research on birds, but who first noticed how long birds mate for? Who first noticed that most birds generally breed as pairs but sometimes break their marriage vows? And who first linked the concept of natural selection to reproductive behaviours such as these?

Big, general concepts and the facts subsequently obtained that show them to be true are much more important than isolated snippets of information. Anyone can collect facts, but few have the vision to

*Aquatic warblers in their extraordinarily protracted sexual embrace.
Painting by David Quinn (1994).*

generalise and seek broad patterns. The topics I discuss in this book are general to all birds and are important aspects of their lives, including sexual reproduction, territory, song acquisition, mating, migration and clutch size. It is precisely because John Ray, in *The Wisdom of God*, used concepts and ideas to unify ornithology that he assumes the central role in my story.

The concepts that underpin ornithology span the entire lives of birds from the moment the egg is fertilised, through hatching, growing up, maturing, acquiring a territory, finding a partner and so on. The 'concepts' here include the notion of 'fertilisation' – who discovered sperm? Who first witnessed a sperm penetrating an egg and starting new life? – and migration: who provided the evidence for migration and, in doing so, dispelled the myth of swallows and other birds over-wintering at the bottom of ponds?

The most obvious way of tracing an idea is to go directly to the oldest sources and see whether it occurs there. If it doesn't, the alternative is to work back through the literature. This is not as easy as it sounds, and is rather like trying to trace the source of a river: as the stream becomes smaller the terrain becomes more rugged, and the stream often divides into smaller and smaller rivulets, making identification of the true source a challenge. As one science historian said:

> The path of science cannot be planned, but is ever tortuous and dendritic, breaking away into innumerable side tracks, which terminate with monotonous and baffling regularity in the wilderness. When the purpose is at length attained, it seems incredible that so much time and energy should have been required and expended to achieve so modest a result.[3]

The further back in time one goes, the harder it becomes to identify priorities. Aristotle, writing in 300 BC, is often considered the root of all biological knowledge, but even he routinely took information from others. Later, in the Middle Ages, naturalists used Aristotle's writings – such as his *History of Animals* – as a kind of scrapbook that they could add to when they saw fit, but without indicating that they were doing so. The result is a huge volume of medieval writings – epitomised by the Bestiaries – in which identifying the origin of any idea is difficult.[4]

Tracing the beginnings of ornithological ideas can be an exciting form of detective work, but because early writers often borrowed unscrupulously from their predecessors it is often tricky.[5] Since my aim is to paint a broad picture of the origins of ornithology, I have followed ideas back as far as possible and used them to identify points of historical and scientific interest that might also stimulate new concepts in ornithology.

John Ray was the turning point. Prior to his *Ornithology* and *Wisdom*, much in the study of birds was confused and fragmented. After Ray, ornithology was never the same again. This book is a journey during which we will assess Ray's remarkable contribution to the study of birds, both by looking back at his predecessors and by looking forwards and exploring the range of topics that comprise modern ornithology.

I started working on this book in the mountains of southern Spain, amid an almost constant procession of birds. It was Easter and migrating birds were streaming across the Strait of Gibraltar, up past the village and over the house: bee-eaters, black kites, harriers, swifts and tiny warblers to name just a few, all winging their way northwards. In the 1750s, John White spent several years on Gibraltar, sending his brother Gilbert in England first-hand descriptions and thus clear evidence for the seasonal movements of birds such as swallows. John White's observations should have dispelled any uncertainty over whether birds migrated or hibernated, but old ideas in ornithology die hard.

Just over five years later, I am back in the same village, writing the last few pages, this time in summer as hurtling hordes of screaming swifts prepare to fly south. Five years may seem like a long time, but, as a university teacher and researcher, getting to grips with the history of birds was squeezed in among many other commitments and would not have been possible at all had it not been for a grant from the Leverhulme Trust. Among other things, their support allowed me to employ Bas van Balen as a research assistant, and I owe him an enormous debt of gratitude for tracking down and translating so many obscure references. I also relied heavily on others for information and assistance, and am indebted to all those who answered my queries or allowed me to quote from their letters or e-mails. I wish to acknowledge all those librarians who patiently helped Bas or myself locate material, but in particular I would like to thank: Clair Castle in the Balfour and Newton libraries, Cambridge, for her help and for allowing me access to the rare books there; Alex Krikellis of the Max Planck Institute ornithology library at Seeweisen; Eleanor MacLean of the BlackerWood Library, McGill University, Montreal; and Ian Dawson at the Royal Society for the Protection of Birds. Special thanks are due to Professor Hans Engländer for allowing me to use his magnificent private library.

I am grateful to those individuals who allowed me to quiz them about their past and who so generously shared their recollections with me, especially: Steve Emlen, Sir Brian Follett, Lord Krebs, Peter Lack, Peter Lake, Bob Montgomerie, Ian Newton, Chris Perrins and Staffan Ulfstrand. I am especially grateful to Lord and Lady Middleton for generously allowing me to reproduce images of Francis Willughby and artwork from the Willoughby collection.

Most people from whom I sought advice or assistance were extraordinarily helpful, but a handful deserve special thanks: Patricia Brekke, Isabelle Charmantier, Mark Cocker, Nick Davies, Nicola Hemmings, Simone Immler, Alison Pearn, Jayne Pellatt, Karl Schulze-Hagen, Rolf Schlenker, Roger Short, Claire Spottiswoode, Dorothy Vincent and Glynn Woods. I am especially grateful to Jürgen Haffer, Linda Hoy and Bob Montgomerie for reading the manuscript and for their constructive comments and suggestions. Any errors are, regrettably, my own. I thank the Leverhulme Trust for their support, encouragement and patience: in a world beleaguered by bureaucracy, their no-nonsense approach to research is a breath of fresh air. My agent Felicity Bryan, and Bill Swainson and his team at Bloomsbury, especially Emily Sweet, provided invaluable inspiration and guidance. Finally, as always, my family Miriam, Nick, Francesca and Laurie deserve the most thanks.

The so-called 'counterfeit' or tapiragem parrot, whose red and yellow feathers were artificially created by the indigenous people of South America, was a cause of consternation for John Ray. This painting of an unknown species by Jacques Barraband is from Levaillant (1801).

1

From Folklore to Facts

John Ray and Ornithology

Imagine a world in which your day-to-day life is ruled not by logic and common sense, but by fear, superstition and a God whose constant succession of 'signs' often appears in the form of birds. Where, seeing a solitary magpie predisposes you to some form of bad luck; where hearing a raven means imminent death; where you can restore a blind person's sight with a special stone found only in the nest of the swallow; whereby placing the semen of a pigeon on someone's dress will make them love you; and where a dead kingfisher suspended from a silk thread acts as a natural weathercock. The magical powers that allowed the kingfisher to anticipate an approaching storm and turn its breast into the wind, meant that it was not uncommon to find their sad corpses in many English and French rural homes.[1]

Such fanciful ideas were very were deeply ingrained[2] and it was only in the 1600s that they began to be challenged. The force for change was Francis Bacon, with his emphasis on experiment and evidence as the basis for reliable knowledge. Another keen advocate of this new way of thinking was the Norwich-based ornithologist Sir Thomas Browne, who also happened to be a friend of John Ray's. Browne's book *Pseudodoxia Epidemica* of 1646 is a gem, championing a new kind of evidence-based science, derived from elegant experiments that put old wives' tales, like the kingfisher story, to the test.[3]

Browne's trial was as simple as it was ingenious: by suspending two kingfishers side by side and showing that they each pointed in a different direction he demolished the myth at a stroke. Browne's attitude epitomised the new approach to natural history and his influence on Ray was immense. As Ray himself later wrote:

> Let it not suffice to be book-learned, to read what others have written and to take on trust more falsehood than truth, but let us ourselves examine things as we have the opportunity, and converse with Nature as well as with books ...[4]

To see what Ray meant by 'book-learned' one need only visit the Newton Library, in the Zoology Department at the University of Cambridge. Concealed inside what can best be described as a huge metal cage at the back of the library is a remarkable collection of special bird books. These volumes, once the possession of Alfred Newton, ornithologist and head of department during the late 1800s, encompass much of the entire history of ornithology. On one particular shelf sit several enormous, leather-bound volumes from the sixteenth and seventeenth centuries whose rarity, fragility and value makes one almost frightened to open them. Arranged in chronological order these are the ornithological encyclopaedias of Pierre Belon, Conrad Gessner, Ulisse Aldrovandi and John Jonston. On the very next shelf lies Ray's own encyclopaedia.

Considering how little was know about birds at that time, why are these books so large, and what on earth did Belon and his fellow authors find to write about? The answer is for the most part bizarre stuff. The content of Gessner's and Aldrovandi's volumes is particularly bewildering. Certainly, there is some information we recognise as ornithology but much of it is buried in a mass of material that today counts as neither ornithology nor natural history and one has to work hard to find it.

These early authors acquired their information from two principle written sources. From Aristotle, the fount of all knowledge, whom they

F. Serinus, Chloridis ueterum species alia.

ITALICE Serin, Scartzerino.

GALLICE Cedrin.

GERMAN. Fädemle/ Schwä=
derle/ Girlitz/ Grill/ Hirngrill.

Chloris Aristotelis, Gaza Luteam
& Luteolam uertit, malim ego
uiridiam.

ITALICE Verdon, Ver=
derro, Verdmontan, Zaranto,
Taranto, & Frinson circa Tri=
dentum.

GALLICE Verdier, Serrant.
Sabaudis Verdeyre: q̃d nomen
etiã passeri spermologo nostro à
Gallis attribui puto.

GER. Grünling/ Grünfinck/
Kuttuogel / Tutter/ Rappuo=
gel/ Hirsuogel.

LAT. Parus maior, Fringillago
Gazæ.

ITAL. Parisola, Parussola, Pa=
risola domesticha, alicubi capo
negro, et circa Alpes tschirnabó.
Priora duo ex his nomina paro=
rum generi communia sunt: Ca=
po negro atricapillæ seu ficedulæ
potius attribuendum est.

GALL. Mesange. Sabaudis
Maienze.

GERM. Spiegelmeiß/ Grosse
meiß/ Brandtmeiß/ Kolmeiß ali
quibus, nostri enim de minore
paro uerticis nigri hoc nomé esse
runt.

Parus cæruleus hic cognominari
potest.

ITAL. Parussolin, Parozolina.

GALL. Marenge.

GER. Blawmeiß/ Bymeisse/
Pimpelmeiß/ Meelmeiß.

*A page from Conrad Gessner's ornithological encyclopaedia of 1555,
whose illustrations – here, hand-coloured – are reasonably accurate,
but whose text is loaded with symbolism and folklore.*

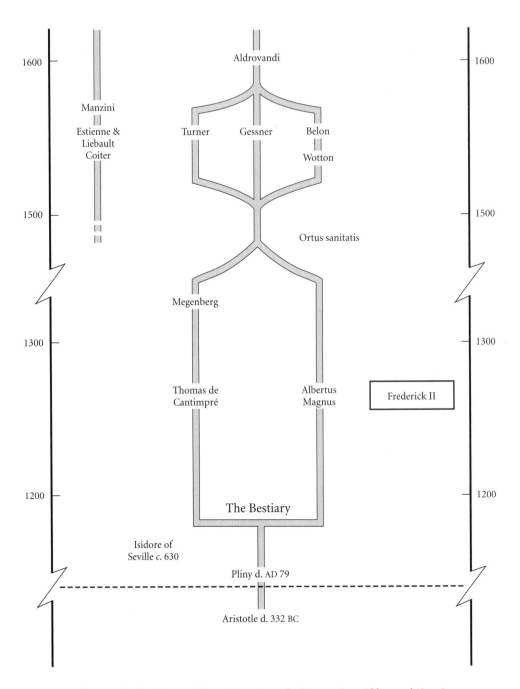

The growth of early ornithology from Aristotle (bottom) to Aldrovandi (top).
Frederick II's writings remained unknown until their rediscovery in 1788,
hence his isolated position here. (Redrawn from White, 1954)

hardly dared contradict; and from a new fashion that was sweeping Europe – emblems, whose own origins lay in the medieval Bestiaries.[5]

Aristotle's works span virtually the entirety of human knowledge, from ethics, poetry and politics to zoology. Described as a one-man university, his success is partly attributable to the absence of ideology or a fixed set of religious beliefs that allowed ideas and scholarship to flourish in ancient Greece.[6] His observations, ideas and ingenious interweaving of philosophy and natural history in the *History of Animals* and the *Generation of Animals* provide the very beginnings of ornithology. He considered birds 'a worthy occupation for the philosophic mind'[7] writing enthusiastically about their reproduction, migration, anatomy, development, territory and taxonomy.

Aristotle's scientific reputation, however, has fluctuated over the centuries. Many, like Charles Darwin, admired him: 'Linnaeus and Cuvier have been my two gods ... but they were mere schoolboys to old Aristotle.'[8] Darwin's endorsement was no guarantee of protection and in the 1980s Nobel prize-winner Peter Medawar and his wife Jean disparagingly described Aristotle's works as a 'tiresome farrago of hearsay'. They singled out his writings on the sexual behaviour of birds, in which he said that some birds like the barn door cock are 'salacious', while others like 'the whole tribe of crows' are 'inclined to chastity', denouncing them as mere gossip.[9] The Medawars, however, were not ornithologists, and their rejection of Aristotle's comment on avian copulation was premature, for paternity studies based on DNA fingerprinting in the 1980s confirmed that what Aristotle had said was true.

For more than a millennium after his death Aristotle's scientific work remained largely unknown. Then in the twelfth century copies of his writings were brought to Western Europe, translated into Latin and suddenly became accessible. They provided the foundation for the medieval animal encyclopaedias known as Bestiaries, whose authors like the Dominican friar and polymath Albert the Great took Aristotle's works and embellished them with observations and comments of their own. A mixture of zoology and Christian moralising, the Bestiaries now seem a curious blend of fact and fantasy, but in their day they were *the* zoological reference books. Animals were valued only for their religious significance and their behaviour interpreted in a moralistic manner, exemplified by the pigeon whose twin offspring represented the love of God and the love of neighbours.[10]

Emblem books of the early sixteenth century were similar to the Bestiaries except that the religious message has been replaced by something more symbolic. The aim, however, was the same: to help the reader understand the world and to lead a good life. An emblem was a kind of puzzle with a moral message and normally comprised three elements: a title, an illustration (often of an animal) and a brief explanation, usually in verse. The title and the image were obscure, and their moral message could be understood only by reading the poem.[11] This emblematic world view is beautifully encapsulated in a witty Renaissance image depicting a peacock – which, according to Pliny, was so vain it could not bear to look at its own ugly feet – staring at its toes, with the motto: *Nosce te ipsum* – Know thyself.

At the time Gessner and Aldrovandi were writing their encyclopaedias in the sixteenth century, to claim to know a particular bird species one had to understand *everything* about it, from the complex web of ancient knowledge, through its folklore, mythology and of course, its emblematic significance. Contrary to what we might think then, the Renaissance encyclopaedias were *not* ornithological textbooks; their purpose was entirely different, which is why they have so often been dismissed as quaint and irrelevant. We can obtain a flavour of what they contained from Gessner's account of the peafowl, which includes everything with a peafowl association; its name in different languages, its habits, the fact that its flesh does not decompose (allegedly) after death; the Peacock River in India; the legend of Argus whose hundred eyes were transformed into the peacock's tail, and so on, including Pliny's statement (based on the male's display posture) that the peacock was ashamed of its feet.

The encyclopaedias on the shelves in the Newton Library span little more than one hundred years but as we move from one to the next

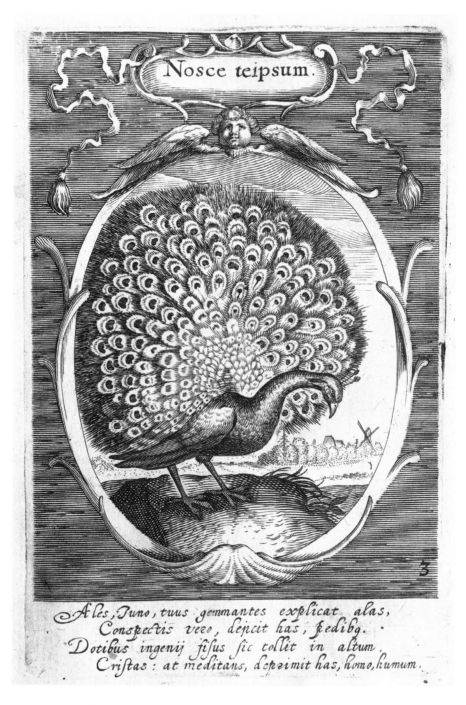

Nosce teipsum.

Ales, Juno, tuus gemmantes explicat alas,
  Conspectis vero, dejicit has, pedibq. .
Dotibus ingenij fisus sic tollit in altum.
  Cristas : at meditans, deprimit has, homo, humum.

A peacock emblem, whose meaning is described in the text. (From Rem, 1617) the amount of emblematic material becomes noticeably less. Gessner's and Aldrovandi's volumes are full of it; Belon's rather less so, and Jonston, who did little more than summarise the others' works, less still. By excluding most of the folklore and stories that make Gessner and Aldrovandi seem so turgid to us, Jonston spearheaded a new era, focusing for the first time almost exclusively on natural history.[12] It was a significant step forward and an approach John Ray and Francis Willughby decided to follow in their own encyclopaedia. In fact they were so keen on this that they spelt it out:

> Having acquainted the reader with our principal aim in this work, which was to give certain characteristic notes of the several kinds, accurately to describe each *species*, and to reduce all to their proper *classes* or *genera*: We shall further add, that we have wholly omitted what we find in other authors concerning *homonymous* and *synonymous* words, or the diverse names of birds, *hieroglyphics*, *emblems*, *morals*, *fables*, *fresages*, or ought else appertaining to *divinity*, *ethics*, *grammar*, or any sort of humane learning: and present him only with what properly relates to their natural history.[13]

What was it about the mid-1600s that encouraged men like Jonston, Willughby and Ray to abandon emblems and folklore in favour of unadulterated natural history? There seem to have been two things. First was the rapid rise in the number of known species. As more and more birds (and other animals) were brought back from the New World they became increasingly difficult to categorise; because they had no symbols or folklore writers could do little else but describe them. Second, in an attempt to wrench society out of its medieval mindset, the sixteenth-century philosopher Francis Bacon denounced the emblematic tradition: the natural world was *not* written in a code revealing the attributes of God; words were arbitrary – they had no hidden meaning, and nor had nature. Instead Bacon argued, true knowledge came from observation and experiment.[14]

*

John Ray was from modest stock; his father was the village blacksmith, his mother a herbal healer. Born in 1627 at Black Notley, Essex, he attended Braintree Grammar School and then – probably supported by a fund organised by the vicar of Braintree for 'hopeful poor scholars' – went up to Catherine Hall, Cambridge, in 1644 at just sixteen. Two years later he moved to Trinity where from 1649 he taught mathematics, Greek and humanities.

Ray had a flair for languages and because of his early classical training was more fluent in Latin than in English and used it for virtually all his scientific works. It made his books more accessible to continental scholars, but was less convenient for those at home who would have preferred him to write in English. Recuperating from an illness during his early thirties, Ray became increasingly fascinated by the local flora. As he travelled round the Cambridgeshire countryside collecting plants he cannot have had any idea at that time how this would lead to a lifetime of monumental botanical industry and unprecedented scholarship.[15]

Francis Willughby came up to Trinity as an undergraduate in 1652 at the age of seventeen. Rich and good-looking, Willughby's interest in natural history was fostered by Ray and the two became firm friends. From an intellectual and privileged background, Willughby became one of the founding members of the Royal Society of London in 1660. Together, he and Ray arranged excursions and during one particular trip to the Isle of Man in 1660 they made the momentous decision to overhaul the entire field of natural history. As Ray's friend William Derham subsequently described it:

> These two gentlemen, finding the history of nature very imperfect ... agreed between themselves ... to reduce the several tribes of things to a method; and to give accurate descriptions of the several species, from a strict view of them, and forasmuch as Mr Willughby's genius lay chiefly to animals, therefore he undertook the birds, beasts, fishes, and insects, as Mr Ray did the vegetables.[16]

John Ray (top), by an unknown artist, and Francis Willughby (probably by Gerard Soest), whose joint efforts produced the first encyclopaedia of enlightened ornithology in 1676.

The plan faltered almost as soon as it had begun. Ray had been ordained – somewhat reluctantly in 1660, the year Charles II was restored to the throne. Two years later, by refusing to sign the Act of Uniformity, Ray's peaceful Cambridge lifestyle was brought to an abrupt end. The Act, designed to purge the Church of its Puritan clergy, required everyone to accept the Book of Common Prayer. As a Puritan Ray refused to sign the oath of loyalty because it compromised his belief and, along with as many as two thousand other clergy, he abandoned the Church. With no source of income he returned to Black Notley and 'cast himself upon Providence and good friends'.[17]

His best friend was Francis Willughby and, rather than being a disaster, Ray's loss of employment was an opportunity which provided him with unparalleled freedom to travel with Willughby (at Willughby's expense) across Britain and continental Europe observing and collecting specimens. In 1662 they visited a number of British islands renowned for seabirds, including Priestholm, Bardsey, Caldey, the Farnes and the Bass Rock, while their European travels took them to Cologne, Frankfurt, Vienna, Padua, and Rome, collecting and dissecting as they went. They bought natural history paintings, visited other naturalists and in 1665 made a special trip to see Aldrovandi's collection of natural history curiosities.[18]

When they returned to England in 1666 after three years of travel, Ray moved in with Willughby at the family seat at Middleton Hall, Warwickshire and together they began to organise their notes and observations. Ray was elected a Fellow of the Royal Society in 1667, and – unusually – was excused the subscription because of his impoverished circumstances. In 1668 Willughby married the heiress Emma Barnard and started a family. The following year, however, he became sick while on a trip to Chester with Ray, marking the beginning of a period of protracted poor health. After a brief respite in 1671, he was taken ill again and died on 3 July 1672 at the age of only thirty-seven. Willughby's widow appointed Ray one of her husband's executors, allowing him to remain in the house in return for educating her two young sons and publishing her husband's manuscripts.

Over the next few years Ray assiduously and intelligently transformed Willughby's incomplete notes into an ornithological encyclopaedia as a memorial to his friend and patron. In 1673 Ray wrote to his friend Martin Lister, saying: 'That the work may not be defective, I intend to take in all the kinds I find in books which Mr Willughby described not, and to have a figure for all the descriptions I can procure for them.' He culled illustrations from a variety of published sources, and Emma Willughby covered the cost of producing new ones. But Ray was disappointed by these images. This is hardly surprising; compared with the fabulous paintings by Johann Walther that Willughby had purchased on their travels, the engravings are poor.[19]

The encyclopaedia, entitled *Ornithologia*, was published with the help of the Royal Society in 1676. This Latin version was not popular however, and Ray soon began an English edition that appeared in 1678: *The Ornithology of Francis Willughby*. Ray added extra material, including sections on falconry, bird catching and keeping, and to make the list of birds as complete as possible added information from the recent accounts of various explorers.[20]

Ray's *Ornithology* was better than anything that had gone before. First, because as Ray himself says:

> It was neither the author's or his design to write pandects [exhaustive treatments] of birds, comprising whatever had before been written by others, whether true or fabulous; *that* having been performed already by Gesner [sic] and Aldrovandus, nor to abridge their bulky volumes, such epitomes having been already made by Jonston.[21]

Second, their objective 'was the formation of a system, and the invention of one which might serve all the requisite purposes, not only in ornithology, but in the several other zoological departments to search for a natural system'.[22] Using his extensive experience of botanical taxonomy, Ray provided the first working definition of a species and then used it to produce the first useful classification of birds.

FRANCISCI WILLUGHBEII

De Middleton in agro Warwicenſi, Armigeri,

E REGIA SOCIETATE,

ORNITHOLOGIÆ
·LIBRI TRES:

In quibus

Aves omnes hactenus cognitæ in methodum naturis ſuis convenientem redactæ accuratè deſcribuntur,

Deſcriptiones Iconibus elegantiſſimis & vivarum Avium ſimillimis, Æri inciſis illuſtrantur.

Totum opus recognovit, digeſſit, ſupplevit

JOANNES RAIUS.

Sumptus in Chalcographos fecit

Illuſtriſſ. D. *EMMA WILLUGHBY*, Vidua.

NVLLIVS IN VERBA

LONDINI;

Impenſis *Joannis Martyn*, Regiæ Societatis Typographi, ad inſigne Campanæ in Cæmeterio D. *Pauli*. MDCLXXVI.

Title page from the original 1676 Latin edition of Ray's Ornithology, sometimes known as the 'widow's edition' because Emma, the name of Willughby's widow, also appears here as a sponsor of the illustrations.

A plate from the 1678 English edition of Ray's encyclopaedia, The Ornithology of Francis Willughby, *showing three European species – tawny owl, nightjar (top) and Alpine swift (lower) – and three unfamiliar ones (middle) redrawn from Marcgrave's* Historiae Naturalis Brasiliae *of 1648.*

Prior to this species had been defined as 'groups of *similar* individuals that are *different* from individuals belonging to other species', but with no special relationship to each other.[23] The notion of 'similarity' was hopeless for those species where the sexes differed or where individuals could occur in different colour forms – like the common buzzard – and worked even less well for organisms like butterflies that metamorphosed through distinct larval and adult stages. The trick was to find a definition that encompassed this variation, and using his botanical experience Ray did exactly that:

> After a long and considerable investigation, no surer criterion for determining species has occurred to me than the distinguishing features that perpetuate themselves in propagation from seed. Thus, no matter what variations occur in the individuals or the species, if they spring from the seed of one and the same plant, they are accidental variations and not such as to distinguish species ... Animals likewise that differ specifically preserve their distinct species permanently; one species never springs from the seed of another nor vice versa.[24]

The fact that two apparently distinct species could interbreed and produce bastard or hybrid offspring was a major challenge to the reality of species. Zoologists had long considered copulations between different species morally unnatural and biologically inconvenient, whereas botanists were much more relaxed about it, possibly because plant hybrids are more common, more passive, and therefore more acceptable.[25]

Straddling both disciplines, Ray is particularly perceptive on this point:

> Birds also of diverse kinds do sometimes couple together, and mingle their seed, from whence proceeds a third and spurious production, which partakes of both kinds; which yet I suppose does not generate its like: for otherwise the number of species in birds would have been ere now almost infinitely increased.[26]

In other words, to be considered as members of the same species two individuals must be capable of producing offspring which are themselves fertile, a definition remarkably close to one we still use today.

When Ray and Willughby were collecting information for their encyclopaedia they came across a story that if true, threatened to undermine Ray's definition of a species. Georg Marcgrave, in his account of the natural history of Brazil, reported how the native people were able to alter the colour of captive parrots at will, using secretions from a poison-dart frog: 'The Tapuia Indians prepare parrots with several colours taking out the feathers and painting their skin with several colours; they are called by the Portuguese … counterfeit parrots.'

Ray didn't quite know what to make of this, since: 'if it be true (for to me indeed it seems not probable) it is to no purpose to distinguish parrots by the diversity of their colour, since therein they may vary infinitely'.[27]

Throughout much of the history of ornithology, the naming and classification of birds had been an ongoing and difficult exercise. Aristotle's attempt involved dividing birds into those living on land, beside water and on water – an ingenious scheme based on a mixture of anatomy and ecology: ducks, for example, had webbed feet and lived on water and were different from crows that lived on land and did not have webbed feet. There was virtually no improvement on this scheme for eight hundred years and even when Belon introduced his system based on habits and habitats that grouped birds into six 'orders', the improvement was minimal. Aldrovandi's arrangement fifty years later, in 1600, was no improvement at all, being based on what he idiosyncratically considered natural groupings: birds that dust-bathed, birds with hard bills (as diverse as parrots, crossbills and raptors) and so on. Gessner was more prudent. He knew that there must be a natural order of birds but, unable to resolve it, he simply presented the birds in his encyclopaedia in alphabetical order.[28]

Ray revolutionised the classification of birds, abandoning Aristotle's system and replacing it with one based mainly on form – such as the shape of the beak or feet – rather than function or habitat. He divided

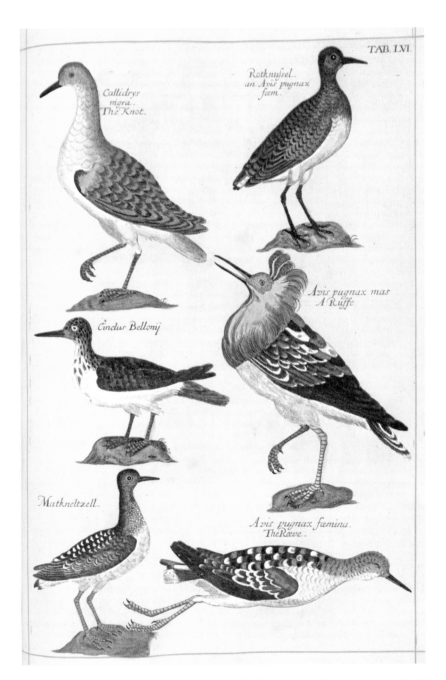

Plate 56 from a unique hand-coloured copy of the English edition of Ray's encyclopaedia, The Ornithology of Francis Willughby, *presented by either Ray or Willughby's widow, Emma, to Samuel Pepys when he was President of the Royal Society between 1684 and 1686.*

birds into land and water species, further sub-dividing them according to the shape of their beak and feet. It was novel, and by and large it worked. Ray's biographer, Charles Raven, shows convincingly that this new system of classification was entirely Ray's doing rather than Willughby's and says: 'No one will claim that he has produced a fully satisfactory classification: no one, who realizes how small were his data and how great the chaos of tradition, will refuse a real admiration for his achievement. He had based his system upon accurate study and dissection; he had tried to take the whole life and structure of each species into account; he has given us the first 'scientific" classification.'[29] Erwin Stresemann, writing in the 1940s, considered Ray's system one of genius, reflecting true evolutionary relationships between birds better even than that introduced by Linnaeus sixty years later.[30]

Ray's practical classification provided the firm foundation that ornithology needed in order to expand. But there was more to the *Ornithology* than classification. By providing accurate, engaging descriptions – refreshingly free of plagiarism – of the birds themselves, Ray and Willughby inspired others to study birds.

The goldcrest:

This is the least of all birds found with us in England, weighing not more than one single drachm ... The top of the head is adorned with a most beautiful bright spot (which they call a crest) of a deep saffron or pale scarlet colour. Hence it got those ambitious titles of *Regulus* [little king] and *Tyrannus* [tyrant]. This crest or crown (if you please so to call it) it can when it lifts, by corrugating its forehead, and drawing the sides of the spot together, wholly conceal and render invisible.

The quail:

Quails are birds of passage: for being impatient of cold, when winter comes they depart out of northern and cold countries into hotter and more southerly; flying even over seas; which one would admire, considering the weight of their bodies and shortness of their wings.

These superb watercolours by Johann Walther, dating from around 1650, are the goldcrest (top left), kingfisher (top right), crossbill (lower left) and shoveller (lower right), and are from a book by Leonard Baltner, purchased by Willughby when he and Ray were in Strasbourg in 1663.

And the cuckoo:

> The *Cuckow* her self builds no nest; but having found the nest of some
> little bird, she either devours or destroys the eggs she there finds, and in
> the room thereof lays one of her own, and so forsakes it. The silly bird
> returning, sits on this egg, hatches it, and with a great deal of care and
> toil broods, feeds and cherishes the young *cuckow* for her own, until it
> be grown up and able to fly and shift for it self. Which thing seems so
> strange, monstrous and absurd, that for my part I cannot sufficiently
> wonder there should be such an example in nature; nor could I have
> ever been induced to believe such a thing had been done by natures
> instinct, had I not with my own eyes seen it. For nature in other things
> is wont constantly to observe one and the same law and order agreeable
> to the highest reason and prudence: Which in this case, is that dams
> make nests for themselves, if it need be, sit upon their own eggs, and
> bring up their young after they are hatched.[31]

A major source of confusion for early ornithologists (in addition to
the counterfeit parrots of Brazil) was the fact that the same species
could not only look completely different depending on its sex, age or
season; in different regions it might also be known by completely different names. For progress to be made it was essential that birds could
be correctly identified in all their different guises and for everyone to
use the same names.

The first attempt to impose some consistency on bird names
occurred in the mid-1500s when Gessner's friend William Turner,
Fellow of Pembroke College, Cambridge, decided to identify all the
birds mentioned by Aristotle and Pliny. Deciphering the ancients'
descriptions and matching them to the birds he knew was a monumental effort but a significant advance for ornithology. Overall,
Turner did a remarkable job, but, because he was the first to attempt
such a task, he inevitably made a few mistakes, confusing the fieldfare
and mistle thrush, and assuming the male and female hen harrier to
be separate species, for example.[32]

Some of the names Turner used, like the *culicilega* (a wagtail) or the *rubicillia* (the bullfinch), have disappeared, while others have become attached to another species, like the junco, which in Turner's day was the reed bunting and is now a North American sparrow; and the *gallinago* then the 'wod cok' (woodcock), today the snipe. Many other names though are uncannily familiar to today's ornithologists who know their scientific names: the *alcedo* (kingfisher); the *certhia* (tree creeper); the *fringilla* (chaffinch); the *merula* (what Turner called the 'blak osel, a blak byrd'); the *pica* (magpie) and the *sitta* (Turner's nutjobber; our nuthatch).

Turner's little book was written in Latin, the scientific language of the day and many of his bird names have their origin either in the 'Latinised' Greek of Aristotle or come from Roman scholars like Marcus Varro (who lived in the first century BC) or Pliny (who lived in the first century AD). The majority of terms incorporated into the scientific names used by ornithologists today, especially for European species, have therefore been around for a very long time.[33]

Linnaeus, born two years after Ray's death, usually gets the credit for arranging and naming organisms and while he certainly invented some bird names he was mainly an organiser, reducing the long, cumbersome names used by his predecessors that also served as descriptions, to his concise binomial system. In the *Ornithology* Ray, for example, referred to the shoveller duck as *Anas platyrhynchos altera sive clypeata Germanis dicta* ('another duck with a broad bill, or according to the Germans, with a shield-like gorget') – hardly convenient. Linnaeus snappily reduced it to *Anas clypeata*. In fact *Anas* can be traced back to Varro; it means to swim, and *clypeata* means shield-bearing. As is now well known, Linnaeus's system comprised a generic name (the genus, in this case *Anas*) and a specific name (the species, here *clypeata*).[34]

There is a nice story of how the puffin acquired its scientific name. With no personal experience of this seabird in his native Zurich, Gessner used a description sent to him by his English friend John Caius. Apparently amused by Caius's description, Gessner wrote: 'If you

Overleaf: *Part of the collection of artwork acquired by Philip II of Spain in the mid-1500s, these paintings of a woodcock and quail are by an unknown artist.*

imagine that this bird was white, and had then put on a black cloak with a cowl, you could give this bird the name of "little friar [brother] of the sea" (fratercula arctica).' Irritated by his friend's frivolity, Caius deleted this phrase from his copy of Gessner's encyclopaedia and replaced it with his own more sensible description. To no avail: Gessner's name stuck.[35]

To ensure that their *Ornithology* was as comprehensive as possible Ray and Willughby included a description of every bird species then known, and charting the number of birds over time provides a revealing measure of how our knowledge has expanded. Aristotle named 140 forms; Thomas de Cantimpré, just 144 in the thirteenth century, reflecting a spectacular lack of progress; by 1555 Belon and Gessner both listed around 200, and in 1676 Ray and Willughby list no fewer than 500. Twenty-five years later when he was writing *The Wisdom of God*, Ray speculated, much as conservationists do today, 'How many of each genus [species] remain yet undiscovered, one cannot certainly nor very nearly conjecture; but we may suppose the whole sum of

The Atlantic puffin, whose scientific name Fratercula arctica, *meaning 'little friar of the sea or arctic', was derived from Conrad Gessner's mental image of a white bird in a monk's cowl. (From Nozeman, 1770–1829)*

beasts and birds to exceed by a third part, and fishes by one half those known.'[36]

In other words, at the end of the seventeenth century Ray thought there might be a further 160 or 170 birds waiting to be discovered, giving a grand total of some 670 species. Over the next century and a half, as exploration continued, the numbers of known birds soared: in 1760 Mathurin Brisson listed 1,500 and a century later Charles Lucien Bonaparte (nephew of the emperor) listed no fewer than 7,000. The increase was not due solely to new discoveries; as the number of birds increased so did confusion and argument over what was and what was not a species. Basing their definition solely on a bird's appearance, museum biologists created hundreds of spurious species, with numbers peaking at almost 19,000 in the early 1900s. By the early 1940s a consensus regarding the biological notion of a species was reached, and numbers dropped to about 8,600. Since then numbers have increased slightly, partly as a result of the dis- covery of some (fewer than two hundred) genuinely new species, but largely because of the re-evaluation of different geographical forms in the light of new information, especially DNA analysis, giving us some 10,000 species today.[37]

Almost all subsequent ornithologists have acknowledged the signifi- cance of Ray's and Willughby's *Ornithology* and many have debated which of the two deserves the credit. Who was the mastermind? Was it Willughby, as John Ray modestly and loyally suggests in the introduction to the English edition? Or was it Ray, as his adulatory biographer John Raven maintains? Both were clearly inspired. By nominating John Ray as the best ornithologist, however, I do not mean to belittle Willughby; had he lived longer we might have seen him in a different light, but Ray's later work consolidated his own position as the superior intellect.

The Wisdom of God is Ray's celebration of the miraculous way every feature of nature, including many ornithological ones, fit so beautiful- ly together. It is an attempt to reconcile natural science and religious

Bird-of-paradise skins were imported into Europe without their legs or feet, giving rise to the notion that, like angels, these birds never alighted. Ray's friend Henry More (1653) was among the first to challenge such notions. (Attributed to Albrecht Dürer)

authority and Ray succeeds both by providing a general explanation for the way things are, but also with remarkable foresight by identifying some of the most fundamental issues in bird biology.

Physico-theology was a new way of thinking about the natural world. Its essence was that God had created a perfect world and provided nature for man's use, often presenting it as a cryptograph; something to be decoded, resolved and interpreted. Occasionally, in His wisdom God provided clues, such as the presence of yellow flowers on plants that could be used to treat jaundice, or red plumage on birds like the crossbill that could cure scarlet fever: the so-called doctrine of signatures. Sometimes, there was no puzzle; God's wisdom was perfectly clear. Swans had long necks to enable them to reach submerged plants and short legs so that they could swim at the same time.[38] In other instances, however, such as the fact that birds lay eggs rather than give birth to live young, God created some real brain-teasers.

The idea of physico-theology originated from two of Ray's Cambridge colleagues, Ralph Cudworth and Henry More. Ray's interest was motivated by the conviction that the great Renaissance thinker René Descartes's view of animals was wrong. One of the main architects of the scientific revolution, the Roman Catholic Descartes considered animals as just so much plumbing – soulless automata. Only man had a soul, and to allow animals anything similar, undermined man's special God-given status. Descartes's view was medieval in outlook; it was a retrograde step and threatened to eliminate the zoological enlightenment that had occurred during the previous fifty or sixty years epitomised by the popularity of the great animal encyclopaedias. To those like the Protestant Ray who knew animals, Descartes's ideas were blatant nonsense and, instead of seeing them as distinct, Ray, Cudworth and More considered all animals to be part of God's great scheme, evidence of His existence and providence.

It is in this spirit that Ray offers a set of reasons for valuing birds:

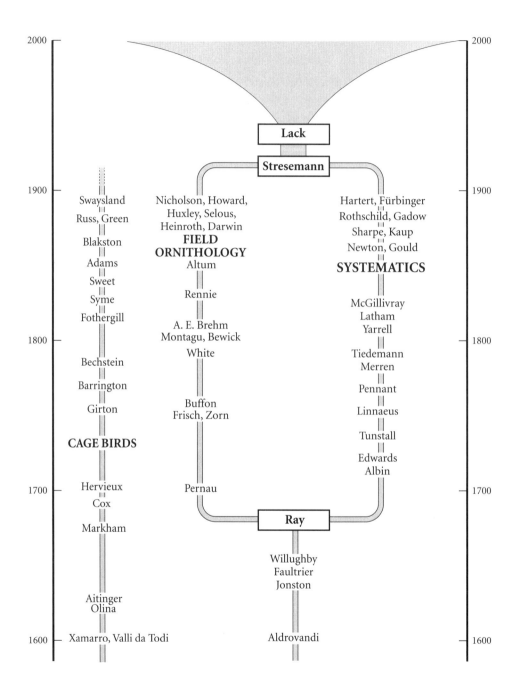

Ornithology from 1600 (bottom) to the 1940s (top). John Ray marks the bifurcation of ornithological interests: his encyclopaedia became the focus for systematics and faunistics, while his Wisdom of God *launched the field study of birds. (Redrawn from Haffer, 2007)*

It cannot be denied that Birds are of great use to us; their flesh according us a good Part of our Food, and that the most delicate too, and their other parts physic, not excepting their very excrements. Their feathers serve to stuff our beds and pillows, yielding us soft and warm lodging, which is no small convenience and comfort to us, especially in these northern parts of the world. Some of them have also been always employed by military men in plumes, to adorn their crests, and render them formidable to their enemies. Their wings and quills are made use of for writing-pens, and to brush and cleanse our rooms, and their furniture. Besides, by their melodious accents they gratify our ears; by their beautiful shapes and colours, they delight our eyes, being very ornamental to the world, and rendering the country where the hedges and woods are full of them, very pleasant and clearly, which without them would be no less lonely and melancholy. Not to mention the exercise, diversion, and recreation, which some of them give us.[39]

There was widespread enthusiasm for physico-theology and some of its proponents managed to see God's wisdom everywhere. At its worst physico-theology was laughable, as in the case where the variegated colour of domesticated animals was considered an example of God's providence in order 'that mankind may more readily distinguish and claim their respective property' or that 'the Creator made the excrement of horses smell sweet', because He knew that men would spend a lot of time in their vicinity.[40] As Richard Mabey has said: 'to sceptical modern minds physico-theology is nothing more than a set of truisms: creatures that were not properly "designed" for their ways of life would simply cease to be'.[41] At its best however, physico-theology helped to identify what we now call adaptations, such as the injury-feigning of a female partridge attempting to protect her brood, or the protective plumage colour of many ground-nesting birds.

Physico-theology also enabled Ray to generalise and to identify biological phenomena he knew required an explanation. Substituting his ecclesiastical interpretation for an evolutionary one, the issues he

This painting was obtained by John Ray during his European
tour with Francis Willughby. The bird is a female pintailed sandgrouse,
a species unfamiliar to Willughby and Ray but a beautiful example of
a camouflaged ground-nesting species.

poses suddenly become extraordinarily modern. Indeed, the questions Ray identifies in *The Wisdom of God* lie at the very core of ornithology and, in fact, biology as a whole. As scientists have long known, progress often depends on asking the right questions and this was one of Ray's great strengths. By recognising what was important he provided a basis for all subsequent ornithological knowledge.

Ray asks, for example, why birds lay eggs rather than giving birth to live young (as bats do); why birds feed their young with food from the mouth rather than suckling them; why birds grow so rapidly compared with mammals. If birds are mere automata, as René Descartes asserted, why do they exhibit such sophisticated behaviour, especially in the care of their offspring? How do birds know at what time in the year to breed? Why are there equal numbers of males and females?[42]

Physico-theology proved extremely popular. Ray's friend William Derham was particularly effective in promoting it in his own version, *Physico-Theology*, of 1713, and it eventually became the basis for the entire English clergyman-naturalist tradition, epitomised by Gilbert White. Across the North Sea in Protestant Netherlands and Germany physico-theology was enthusiastically embraced in the early 1700s, by among others the nobleman Baron von Pernau and the clergyman Johann Zorn, both of whom made important contributions to ornithology. By studying live birds, both in captivity and in the field, Pernau and Zorn were ornithological innovators. The fashion for keeping songbirds in particular increased around this time and the early 1700s saw too the publication of a number of bird-keeping handbooks, most of them published anonymously (thereby helping to maintain their obscurity) but none the less full of remarkable bird biology.[43] Zorn's own book, inspired by the idea of physico-theology, was named *Petino-Theologie*, meaning winged-theology, or orni-theology.[44]

In the early 1800s physico-theology was given a further boost by William Paley, whose hugely successful *Natural Theology* was largely plagiarised from Ray's *The Wisdom of God*. Paley is now best known for his parable of the watch. Imagine finding a watch, he says; just looking at its intricate design tells you that it must have a designer. Now look at nature; once again, the ingenious design, the intricate fit between an organism and its environment tells you that it must have had its designer, and that designer is God.

Among those inspired by Paley's book was Charles Darwin, who, had he completed his undergraduate studies at Cambridge instead of setting off around the world and subsequently changing it, would also have become a typical clergyman-naturalist. In a way he was one, quietly closeted away at home, except that for him natural selection provided a better explanation for natural history than God. Indeed, with its focus on adaptation, natural theology prepared the ground for Darwin's evolutionary ideas, culminating with *On the Origin of Species* in 1859.

Conceived during his round-the-world voyage on the *Beagle* in the 1830s, natural selection had a long gestation. It was developed and refined over the following twenty years but not presented to the public until Alfred Russel Wallace independently came up with the same idea in 1858, forcing Darwin to reveal his hand. For Darwin natural selection emerged from a combination of several disparate observations, including the ability of breeders to capture and exaggerate some of the natural variation exhibited by animals and plants through selective breeding. The existence of variation – in size, shape, colour and behaviour – and the fact that these traits are often heritable, were two key parts in the theory. The third component, which came to Darwin after he had read Thomas Malthus's essay on population in 1838, was that despite the tremendous potential for organisms to increase, their numbers appear to remain similar, implying that many must die. Darwin put two and two together and reasoned that the environment unwittingly acted like an animal breeder, selectively eliminating those individuals with the least appropriate traits – the less fit – leaving the best adapted to survive and propagate their kind.

As far as Darwin was concerned, God was unnecessary: 'The old argument from design in nature, as given by Paley [and of course Ray], which formerly seemed to me so conclusive, fails, now that the law of natural selection has been discovered.'[45] The exquisite match between an organism and its environment – *adaptation* – was the outcome of the cumulative effects of a purely mechanical process – *natural selection* – that beautifully accounted for the transmutation or *evolution* of species over time. For Darwin and his numerous followers the only process that can explain good design in nature is natural selection.

Ray's intellectual endeavours generated two types of ornithologist: the systematist, whose interests lay in classification and taxonomy, and the field ornithologist who was interested in the behaviour and ecology of birds. For more than two hundred years after Ray's death the systematists regarded themselves as the only true ornithologists. They were the professionals, the *scientific* ornithologists. They considered the field ornithologists with their physico-theology roots to be dilettantes and amateurs – which they were in some ways since many were clergymen or teachers – whose work was worthless. By the late 1700s the two areas of ornithology were epitomised by their respective champions. On the systematists' side was Linnaeus, arrogantly convinced that he alone had been chosen by God to de-code His natural system. A purist, Linnaeus made no attempt to popularise his writings. On the other side was Georges-Louis Leclerc, Comte de Buffon, who wrote about birds in a clear and often exciting way in nicely illustrated books available in variously priced editions. Buffon's cavalier attitude to classification incensed Linnaeus, who got his own back by giving a particularly foul-smelling plant the name *Buffonia*.[46]

The intellectual crevasse that separated systematics from field ornithology was wide, deep and persistent. Writing in the early twentieth century historian and ornithologist William Mullens described the museum men as 'united in a common hatred and contempt for the field-naturalists'.[47] It is particularly ironic that the museum men despised the field men so completely, for it was the latter who finally resolved the fundamental problem of what constituted a species. In fact, ornithologists like Pernau and Zorn, familiar with the ecology, behaviour and appearance of birds, were using a biological concept

THE
WISDOM
OF
GOD

Manifested in the

WORKS

OF THE

Creation.

BY

JOHN RAY, M. A.

Sometime Fellow of Trinity-College
in Cambridge, and now of the
Royal Society.

LONDON:

Printed for Samuel Smith, at the Princes
Arms in S. Pauls Church-Yard, 1691.

Title page of the first edition of John Ray's The Wisdom of God.

of the species from the early 1700s. The museum buffs on the other hand, restricted to bird skins and skeletons, struggled for another two centuries to understand what a species was.[48]

It all finally came together in the 1920s as a result of a particularly remarkable, young German ornithologist named Erwin Stresemann. By expanding the boundaries of traditional museum-based ornithology to embrace the study of wild birds, Stresemann drew the two disparate strands of bird study together to create a new ornithology. By recognising that birds were eminently suitable for investigating areas of biology as diverse as physiology, functional morphology, ecology and behaviour, Stresemann, still only in his twenties, revolutionised ornithology, making it scientifically respectable and part of mainstream zoology.[49]

The great evolutionary biologist and ornithologist Ernst Mayr later wrote: 'No one in the last 100 years has had as profound an impact on world ornithology as Erwin Stresemann.'[50] Despite Mayr's eulogy, Stresemann remains little known outside his native Germany. In the meantime, still uncertain about whether or not the chicken came first, we will start our story of ornithology with the egg.

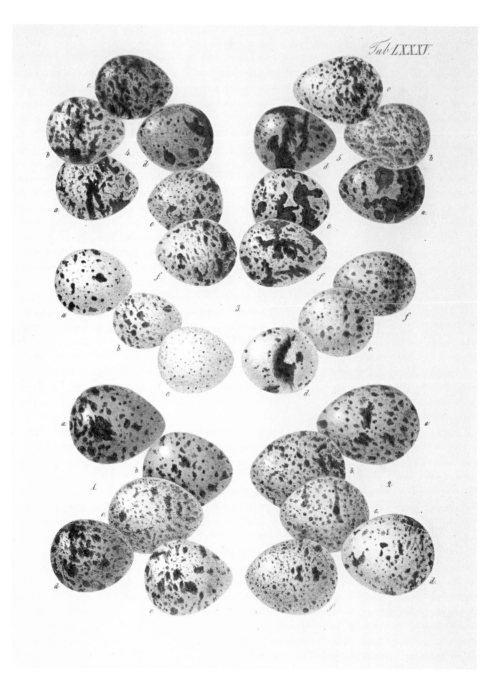

*Birds' eggs are the ultimate external embryo-development system.
The beautifully coloured shells provide protection and in some cases allow
females to identify their own eggs. (From Thienemann, 1845–54)*

2

Seeing and Not Believing
From Egg to Chick

T he hen's eggs we eat are usually infertile. Imprisoned in their battery cages, commercial layers have probably never even seen a cockerel let alone enjoyed the brief, but often brutal copulation essential for impregnation. In earlier times cockerels and hens roamed freely together in the farmyard, the cockerel's irrepressible lust a conspicuous reminder of what life was all about. Yet, despite knowing that the male delivered a teaspoonful of hot semen to the hen, how this resulted in the generation of new life was one of the most perplexing mysteries in the history of biology.

Particularly puzzling was the fact that the contents of newly laid eggs produced by mated hens were utterly indistinguishable from those laid by hens in the absence of a male and therefore infertile. On the face of it, the male's semen seemed redundant for it appeared to add nothing whatsoever to the egg.

With the benefit of considerably more knowledge and some wonderful technology unavailable in John Ray's time, we can now take a freshly laid egg and establish very easily whether it has been fertilised or not. This is what my research students and I do in my laboratory, so allow me to demonstrate.

In front of you on the bench is a microscope, a small plastic Petri dish of saline solution, some finely pointed forceps, a sharp pair of scissors and a fresh hen's egg of unknown provenance. I crack open the egg and gently pour its contents into the dish. This is such a familiar sight that we rarely stop to consider what an incredible design a birds' egg is: a self-contained embryo-development system – a truly wonderful adaptation.

The yellow portion, which in everyday language we call the 'yolk', is in reality a single, enormous cell more appropriately referred to as an 'ovum'. Like all cells, the ovum contains a nucleus, which in birds' eggs sits inside a milky-white spot 2 or 3 mm in diameter on the yolk. This is the business part of the egg – technically known as the germinal disc – and it contains the DNA and the genetic wherewithal to produce a new individual. The rest of the ovum is yolk, a luscious mixture of fat and protein, poised to fuel a burst of extraordinarily rapid embryo growth. The ovum's spherical shape is maintained by a gossamer-thin layer of tissue whose presence becomes obvious only if it is punctured. This easily overlooked tissue is what we are about to examine.

But before we do, let's just consider the white, or albumen, a watery mixture made glutinous by the addition of proteins. The albumen's role is largely to protect the ovum from damage by acting first as an all-round cushion, and second by means of two specially twisted strands that hold it in position. Known as the chalazae (from the Greek *chalaza*, meaning a knot), these strands will be familiar to you in cooked scrambled eggs where they often appear as unappetising, gelatinous nodules. Not only do the chalazae suspend the ovum in the protective albumen; more crucially, they allow the ovum to turn so that in a fertile egg the embryo remains uppermost in the egg even as it is turned, and not fighting gravity as it grows.

With a practised hand I cut the ovum in half using the scissors, quickly grasping each side with the forceps as the yolk oozes out across the dish. The yolk drains away, leaving two tiny globs of grey tissue on the tips of my forceps. As I rinse them in the saline, they reveal their hemispherical form: two halves of a bag that once contained the yolk. Cutting a portion of the tissue from the area that had lain over the germinal disc and washing it again in the saline, I can separate it

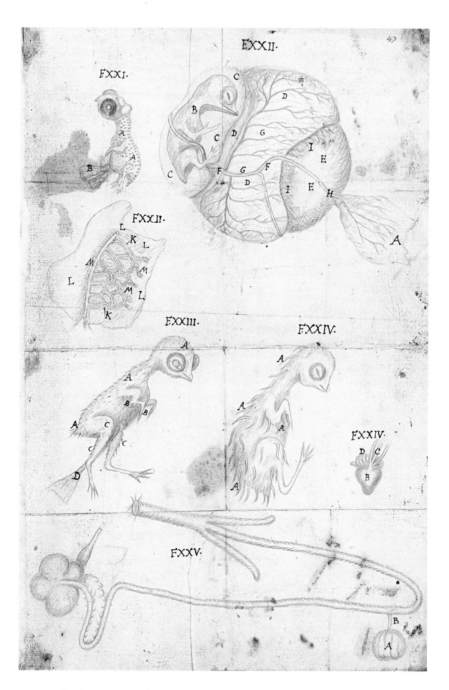

The developing embryo of the fowl. This drawing in red chalk is by the seventeenth-century Italian anatomist Marcello Malpighi, who was the first to recognise that the germinal disc of a bird's egg contains the female nucleus.

into an inner and outer layer, rather like peeling a sticky label off its backing sheet. Keeping track of which is which, I carefully spread both layers flat on their own glass microscope slide.

Taking the slide with the outer layer, I add a drop of fluorescent dye and place it on the microscope stage. As my eyes adjust to the darkness, I gradually become aware of dozens of identical structures, each one an electric-blue crescent moon on a black night sky. These are the fluorescently stained nuclei of sperm trapped on the ovum's outer layer. Adjusting the light I can see more details, including the head of each sperm and their hair-like tail. For these particular sperm this was as close as they got to the female DNA: their journey ended here.

Next, I examine the second slide, the inner layer. Another night sky under the microscope, but this is different – no sperm now, just a cluster of ghostly black holes, the signature of the twenty or so sperm that successfully burrowed their way towards the female nucleus.

Everything we have seen so far – the large numbers of sperm on the outer layer, the holes in the inner layer – suggests that the egg was fertile, but to be sure I examine a speck of material from the germinal disc. It is another visual extravaganza: the thousands upon thousands of electric-blue oval shapes I see under the microscope are the nuclei of cells, the ultimate confirmation that one of those sperm that penetrated the inner layer found and fused with the female DNA, triggering a cascade of cell division as a new life begins. Had the egg not been fertilised there would be no cluster of blue nuclei.

For me, examining eggs in this way is routine, so routine in fact that I sometimes forget just how perplexing and frustrating it must have been for my scientific predecessors as they struggled to piece together the story of fertilisation.

The origin of new life, or 'generation' as it was then known, was for John Ray the single most important issue in the study of natural history: 'The greatest of all the particular phenomena is the formation and organisation of the bodies of animals, consisting of such variety and curiosity.'[1] In the seventeenth century the term 'generation' had a double meaning as Ray indicates by referring to both the 'formation' and 'organisation' of bodies: conception and the development of the embryo, respectively. Even though Ray recognised that these were topics way beyond his abilities, he was able nevertheless to pose some pertinent questions, asking, for example: Where does new life originate? Why do the embryos of birds develop so rapidly compared with mammals? Why do birds, and birds alone, produce large, yolky hard-shelled eggs? Why don't they give birth to live young and suckle them, like bats and other mammals? Ray's cautious responses to these questions inevitably involve the intelligent designer, but despite this still reveal an extraordinary biological insight.[2]

In essence Ray identified three fundamental questions relating to the lives of birds: (i) What is the origin of new life, that is, the basis of conception? (ii) Why do birds lay hard-shelled eggs rather than give birth to live young? (iii) How does new life develop; how do embryos develop, are they preformed or constructed?

In fact, it was the quest of many naturalists both before and after Ray to answer these questions and the only way to do so was to disentangle the bewildering relationship between eggs, copulation, fertility and embryo development. The domestic fowl or chicken was the perfect study species: it was abundant, tame and reproduced virtually throughout the year. It was also well known since ancient times that the breeding biology of chickens was strikingly different from both our own and that of other mammals in one very important way. From the time of Aristotle and probably long before, it was common practice for women (who did not own a cockerel) to take a hen to a neighbour's cockerel to be mated, knowing that after returning home the hen would lay fertile eggs for several weeks!

One of Ray's predecessors, the great seventeenth-century Italian anatomist Hieronymus Fabricius de Aquapendente (Fabricius for short), was among the first to try to understand the prolonged fertility of hens by resolving the links between copulation and fertilisation. His starting point was the notion that hens could lay fertile eggs for a full year after copulating with a cockerel. This was an error, but perhaps an understandable one at the time, and it is possible that Fabricius

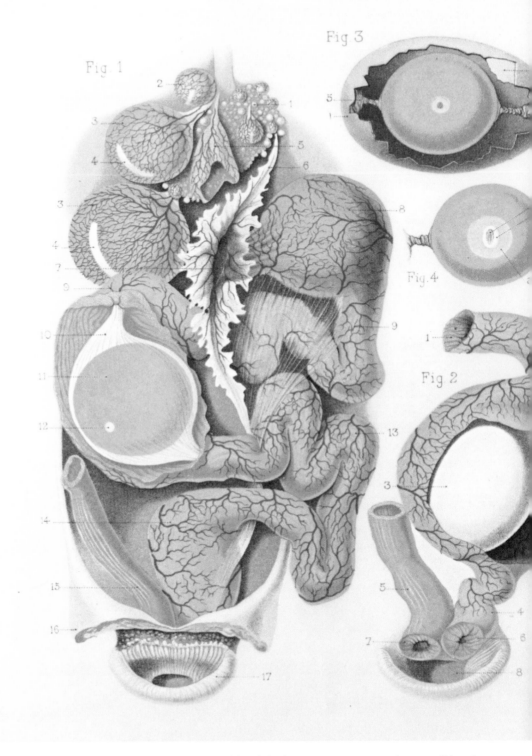

Fig. 1

Fig 3

Fig. 4

Fig. 2

The different regions of the reproductive tract of female birds. In Fig. 1 we can see the ovary (1) with its hierarchy of developing ova (2–8); the infundibulum (7); the oviduct (9, 13 and 14); and an ovum (11) in the uterus (shell gland), with the white (albumen, 10) starting to be formed and the germinal disc (12) on the ovum. In Fig. 2 we see a fully formed white-shelled egg in the uterus; 4 is the vagina

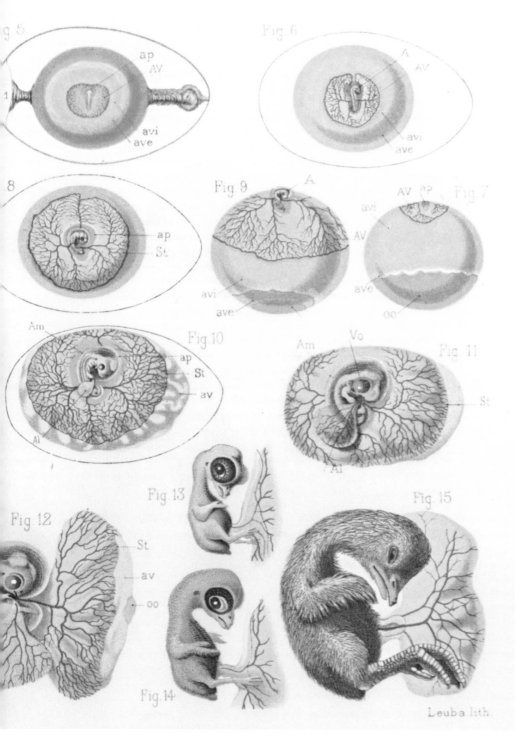

(the sperm storage tubules lie just above this towards the uterus – see the illustration on page 86); 5 is
the gut; and the entrance to the oviduct (6) and the anus (7) open into the cloaca (8). In Fig. 3 we see
a freshly laid egg with the ovum suspended between the chalazae; we also see the germinal disc (the pale
spot) in which the primitive streak – the developing embryo – will appear (see Figs 5 and 6).

did not literally mean one calendar year, but rather a 'season'; that is, late spring and summer, which is more reasonable, but still too long. It wasn't until several years later, when Fabricius's student William Harvey (of circulation fame) took up the problem and conducted his own experiments, that the true interval of three weeks between copulation and the last fertile egg was established.[3]

In a sense it didn't matter that Fabricius believed the interval to be much longer; the three-week delay between the last copulation and the production of the last fertile egg was still extraordinary and cried out for an explanation. Fabricius had two somewhat contradictory ideas. On the one hand he assumed that hens stored semen for weeks after mating and used it to fertilise each egg as it was laid. On the other hand, he also assumed that an insemination would fertilise the ovary's entire complement of ova simultaneously. As Fabricius knew, the hen's ovary consists of a mass of 'eggs' or, more correctly, ova,[*] and if all these were fertilised (he used the lovely term 'fecundated') simultaneously this would account for the protracted period of fertility.[4]

In support of his first idea, Fabricius claimed to have found the precise point where females stored the male's semen – a tiny blind-ending cavity within her cloaca (the common opening of the reproductive and digestive tracts). William Harvey, always sceptical of his mentor's ideas, thought this unlikely, and, indeed, when he checked he could find no trace of semen inside it. What is more, and as Harvey smugly pointed out, Fabricius overlooked the fact that males possess an identical little cavity in their cloaca making it even more unlikely that this was a semen store.[5]

As far as Fabricius's second idea was concerned, he had absolutely no idea of how semen 'fecundated' eggs. He could find no trace of semen inside the female, and could only assume that it acted in an 'ethereal' manner.[6]

What Fabricius started, Harvey tried to finish. After resolving the circulation of blood, Harvey switched his focus to reproduction. Like

* i.e. the yolk with its nucleus – they cannot be called eggs until the white and shell have been added.

Fabricius, one of his goals was to figure out the role of the semen, and to this end Harvey did the obvious thing: he dissected females soon after they had copulated. He dissected both chickens and deer. As royal physician Harvey had unprecedented access to the king's deer to the extent that matings were arranged, and the females summarily dispatched immediately afterwards to satisfy Harvey's reproductive curiosity. But both hens and hinds died in vain, for in neither case could Harvey find any semen. Thwarted, he wrote:

> There is nothing to be found in the uterus after coition than there was before ... the seed of the male after his coupling with the female, does not remain in her womb in any sensible bulk: but (as it seems), evaporates and incorporates itself, either into the body of the womb, or rather into some more interior part.[7]

Harvey speculated that even if there was no semen visible in the oviduct after mating, it should be visible inside the egg if it played a role in generation. Yet the internal appearance of fertile and infertile eggs at the time of laying was identical. Three hundred years earlier, Albert the Great had generated the silly idea that the chalazae were the male's semen. The chalazae do indeed resemble semen in appearance, but they resemble human semen much more than fowl semen – which probably says as much about the way the medieval mind worked as anything else. But Albert added insult to injury by declaring chalazae to be absent from infertile eggs. Harvey dismisses Albert's claim as nonsense, pointing out that the chalazae – which he calls 'treddles' – are identical in fertile and infertile eggs:

> These treddles are found in all the eggs of all birds, as well subventaneous* as fecund. Whence appears the common mistake of our housewives, who think that the treddles are the cocks sperm, and that the chicken is formed of them.[8]

* Subventaneous = infertile.

Ray and Willughby included this quote from Harvey in their *Ornithology*, but added pointedly: 'This is a mistake not of old women or common people only, but also of great physicians and naturalists ...', presumably referring to Aldrovandi, who as late as 1600 was still claiming that the chalazae were semen.[9]

Despite his careful, systematic observations Harvey signally failed to establish the role of the male in reproduction. It was obvious that semen was transferred to the female during copulation and was essential for fertilisation. The problem was that immediately after insemination the semen seemed to disappear. Reluctantly, but with no alternative, Harvey reverted to Fabricius's idea that semen must act in an 'ethereal' manner, like contagion in disease, changing things without touching them: action at a distance. It was the only thing consistent with his observations, but on reading Harvey's account one has the strong impression that he knew it was wrong.[10]

Utterly frustrated by his inability to resolve this particular issue, Harvey abandoned his manuscript and notes for the book he was planning. A further forty years were to pass before he was persuaded to hand over the bundle of papers to his friend George Ent, who, recognising its value, dutifully put the material together and published the *Disputations Touching the Generation of Animals* in 1651, by which time Harvey was seventy-three.

The main conclusion from this wonderful – if incomplete – compendium of reproductive observations, experimentation and careful thought is that it is the egg rather than the semen that plays the primary role in reproduction. Whatever it is that semen does, Harvey says, because it acts in an ethereal manner it adds nothing material to the developing embryo. So convinced was he of the egg's overarching importance that he advertised the fact by printing the phrase *Ex ovo omnia* (all things from the egg) on the title page of his book.[11]

Ray and Willughby quoted extensively from the *Disputations*, keen to include the most up-to-date information in their encyclopaedia. Harvey's idea that the egg was the key to reproduction also coincided with their own observations based on dissections of both male and female birds in breeding condition. Writing in their *Ornithology* they say:

> All animals come from eggs, as well as those called viviparous [live-bearing] as oviparous [egg-laying]: For the females of viviparous have eggs within them, though they do not bring them forth. To wit, those two bodies, commonly called female testicles, are nothing else ... but knots or masses of very small eggs, as will manifestly appear to anyone that shall dissect them; so that we cannot but wonder that a thing so plain and evident should so long escape the observation of the curious and inquisitive eyes of ancient and modern anatomists ... Yea if we consider the matter more exactly, we shall I think find, that the seeds or eggs of viviparous creatures do indeed answer to the cicatricule of eggs, in which from the beginning the young is included.[12]

The cicatricule is the germinal disc and Ray and Willughby are referring here to the very recent discovery – using a microscope – by the Italian anatomist Marcello Malpighi that the germinal disc contains a nucleus. Malpighi's discovery added extra weight to those philosophers – referred to as ovists – who, like Harvey, believed the egg to be all-important, since this nucleus was obviously the germ of the new organism.[13]

This crucial piece of the reproductive puzzle came too late for Harvey, who died in 1657, but, had he known it, he would have been pleased for although it did nothing to resolve the issue of semen, it confirmed his belief in the egg's primacy. It is also possible that Malpighi's discovery might have alerted him to the value of using the new technology. Harvey had been wonderfully astute, but not farsighted enough to use a microscope, and without one he had no hope of either locating or understanding the role of the semen. Microscopes in one form or another had been available since the 1590s so Harvey had no real excuse for not using one. Nor had Ray and Willughby, especially since Robert Hooke's book *Micrographia* illustrating the wonders of the microscope had caused a minor sensation in London when it appeared in 1665. Yet, as far as we know, neither Ray nor Willughby ever used one, presumably because they felt that the microscope had little relevance to the study of birds.[14]

Just a few years after their *Ornithology* appeared there was a major shift in focus when, in 1679, the Royal Society published Antonie van Leeuwenhoek's letter describing the existence of 'animalcules' in semen. Observing them in his own semen initially, but later in a range of other species, including cockerels, Leeuwenhoek made the bold and brilliant suggestion that these 'animalcules' combined with an egg to create a new organism.[15]

Willughby had died in 1672 and so was unaware of Leeuwenhoek's discoveries, but Ray could not ignore them and it was in his book on mammals published in 1693 that he decided to confront Leeuwenhoek's animalcules.[16] By that date, animalcules were only one of several major reproductive problems taxing the minds of philosophers and

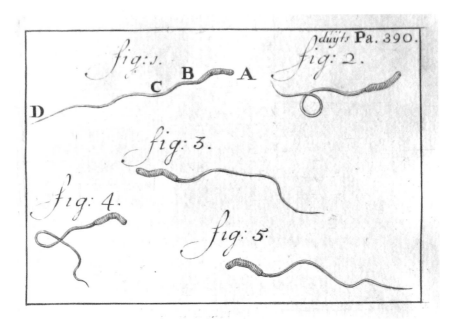

*The first sightings of bird sperm – those of a rooster – as seen by
Antonie van Leeuwenhoek in the 1670s through one of his microscopes,
whose single lens he made from a drop of molten glass.*

scientists. The others included the issue of spontaneous generation and the question of whether individual animals are produced anew each generation.

The idea of spontaneous generation goes back to at least the first half of the fifth century BC, when the Greek philosopher Anaximander claimed living things to be formed out of inorganic matter by the effects of solar heat. Anaximander also stated that the embryo similarly arose from the heat of the mother 'irradiating' the semen of the father. Two centuries later Aristotle also assumed that this was the basis for the origin of new life, particularly for organisms like maggots or flies that appeared 'spontaneously' from rotting meat. With no better alternative, spontaneous generation remained the standard explanation for most medieval scholars, and – incredibly – was not finally put to rest until the nineteenth century. The problem, of course, was that all life has its origin in microscopic structures that are impossible to visualise without the appropriate technology. Following the authority of Aristotle, men as learned as Isaac Newton, William Harvey and René Descartes continued to believe in spontaneous generation.

Ray, however, was in no doubt that spontaneous generation was wrong: 'I entirely agree with those who think that no animal at all is born spontaneously.' He gives several logical reasons, including the existence of two sexes and their elaborate reproductive apparatus, which would be superfluous, he says, if they did not need to bring them together for reproduction. If an insect can be produced spontaneously, he asks, why not an elephant – or a man?

Ray is particularly critical of what he calls the 'mechanical philosophers' like Descartes:

These mechanic philosophers being no way able to give an account thereof from the necessary motion of matter, unguided by mind for ends, prudently therefore break off their system there when they should come to animals, and so leave it altogether untouched. We acknowledge indeed there is a posthumous piece extant, imputed to Cartes [Descartes], and entitled, *De la formation du Foetus*, wherein there is some pretence made to salve all this fortuitous mechanism. But as the theory thereof is built wholly upon a false supposition, sufficiently confuted by our Harvey in his book of *Generation*, that the seed doth materially enter into the composition of the egg ... I shall only add, that natural philosophers, when they endeavour to give an account of any of the works of nature by preconceived principles of their own, are for the most part grossly mistaken and confuted by experience.[17]

The second issue – whether individuals are produced anew each generation – was trickier. The traditional standpoint was that God had made everything, while Ray's more enlightened view was that while He might have done so originally, He also made animals and plants responsible for their own reproduction. To confirm this, Ray calls on his own observations of the ovaries of birds that he noticed consisted of a multitude of little ova ready to be developed, and which he (correctly) assumed represent the female's lifetime supply.[18]

Addressing the question of animalcules and the issue of which sex plays the decisive role in reproduction, Ray decides that the sperm and ovum probably fuse to form a new individual. This isn't quite as momentous a revelation as it might at first seem and was probably a compromise, accommodating both Malpighi's discovery of the nucleus within the germinal disc and Leeuwenhoek's discovery of animalcules. As Ray himself acknowledged, the whole business of generation was 'a subject too difficult for me to handle' and he leaves open the question of spermatozoa:

> How far the animalcules observed in the seed of males, may contribute to generation, I leave to the more sagacious philosophers to enquire, and shall here content myself with referring the reader to the several letters published by Mr. Lewenhoek [sic].[19]

In truth, Ray was not overly enthusiastic about the idea of sperm playing a vital role in reproduction. There were simply too many of them: a provident God would not be so profligate:

The new opinion of Mr Lewenhoek [sic] ... I am less inclinable to, because of the necessary loss of a multitude, I might say, infinity, of them, which seems not agreeable to the wisdom and providence of Nature ... For supposing every male hath in him all the animalcules that he shall or may eject; they may, for ought I know, amount to millions and millions, and so the greatest part of them must needs be lost ... But if we suppose the foetus to be originally in the egg, it is not so.[20]

It would be another one hundred and fifty years before the intimate relationship between eggs, copulation and fertilisation was resolved.

The question of why birds lay large, hard-shelled eggs rather than giving birth to live young was first raised by Ray's friend Henry More in his *Antidote against Aetheism*.[21] A keen proponent of physico-theology, More proposes that laying eggs rather than producing live young is God's way of allowing birds to rear more offspring simultaneously. If, like bats, they produced only one or two at a time they might be feeding young all year. In *The Wisdom of God*, Ray takes up the argument and develops it somewhat more pragmatically, saying:

For if they had been viviparous, the burden of their womb, if they had brought forth any competent number at a time, had been so great and heavy, that their wings would have failed them, and they become an easy prey to their enemies: Or, if they had brought but one or two at a time, they would have been troubled all the year long with feeding their young, or bearing them in their womb.[22]

The idea that producing eggs one at a time (rather than bearing live young) is a weight-saving device and therefore an adaptation for flight has remained the favourite explanation since Ray's time, but a moment's thought renders it unconvincing. Flightless birds exist in no fewer than fifteen different bird families, yet not one of them has given up eggs in favour of bearing live young. Moreover, bats give birth to young that may weigh as much as 40 per cent of the female's weight, whereas (individually)

Once considered a kind of bird, the bat, like all mammals, gives birth to live young. Most bats produce only a single offspring at a time, but some, like this (unidentified) species, produce two. Why birds never evolved the ability to bear live young has been a long-standing mystery. (From Jonston, 1657)

the eggs of most birds represent less than 12 per cent of female weight. In addition, it is also clear from the fact that some birds double their body mass with fat prior to setting off on migration that certain species have the capacity to carry more weight than they typically do.

There have been several ideas as to why live-bearing has not evolved in birds. The first is that tucked up inside its shell *and* within a uterus an embryo would find it difficult to obtain enough oxygen in this 'double layer'. The second argument is based upon the fact that the highly specialised lungs of birds require a period just prior to hatching during which air breathing through the lungs and respiration via the embryonic tissues occur simultaneously – something that is apparently possible only in an egg. The third argument – and this is a bit more esoteric – concerns the sex chromosomes. In mammals, only males carry the distinctive, sex-determining Y chromosome, and it is this and its androgen-stimulating genes that male embryos use to neutralise the feminising effect of maternal hormones they encounter *in utero*. This isn't an option available to birds because it is the female that carries the sex-determining Z chromosome, so the only way to protect the developing male embryo from dangerous maternal hormones is for it to develop outside the female's body, inside a hard-shelled egg.

None of these explanations is totally convincing.[23] Moreover, they all presuppose, somewhat arrogantly, that the birds' system of embryo development is inferior to that adopted by mammals. The various ideas for why birds lay hard-shelled eggs also imply that birds have had the evolutionary opportunity to break free from their reptilian ancestry and abandon egg-laying in favour of live birth. But live birth would work in birds only if they had the ability to lactate, and, as Henry More points out, 'Birds have no breasts, young stay long in shell and then are fed by mouth by parents – observations that plainly signify that Nature does nothing ineptly.'[24]

Perhaps the most plausible idea of all for why birds lay hard-shelled eggs rather than produce live young is linked with their body temperature. At around 41°C the temperature of birds is several degrees higher than that of most mammals and may simply be too high for normal embryo development. As Ray noted, embryonic development in birds is extraordinarily rapid anyway, and we now recognise that pushing it even further through a warmer, internal environment could be disastrous. The eggs of most birds are incubated at around 37°C, a few degrees below body temperature, and with good reason. Experiments by poultry biologists show that eggs incubated at temperatures greater than this develop too rapidly, resulting in embryo death.[25] Their high body temperature therefore appears to answer Ray's question as to why birds do not give birth to live young.

Birds' eggs provide a wonderful window into the extraordinary and rapid changes that take place as the embryo develops. It is a phenomenon that has inspired awe and wonder from the fifth century BC onwards:

Take twenty eggs or more, and set them for brooding under two or more hens. Then on each day of incubation from the second to the last, that of hatching, remove one egg and open it for examination. You will find that everything agrees with what I have said, to the extent that the nature of a bird ought to be compared with that of man.[26]

This statement, loosely attributed to Hippocrates, who, born in 460 BC, is often considered the founder of medical science, is profound in two ways: it recognises the fundamental similarities in the ways birds and humans develop, but it also endorses a *systematic* approach to the study of embryos – the day-by-day examination of the chick's development. Unfortunately, this advice, which held the key to understanding how a hen's egg is transformed from 'nothing' to a fluffy, self-feeding little bird in just three weeks, was ignored for more than two thousand years.

Hippocrates himself certainly never followed this advice, for how else could he have asserted that everything in the embryo is formed simultaneously? More correctly, he suggested that the yolk gave rise to the chick, which was nourished by the white. A century later, Aristotle disagreed on both points. But he too failed to make systematic observations, merely noting that blood was visible after one day, the heart on the third and by the tenth day all organs were discernible. He reported that at hatching the yolk is still attached to the chick's gut and how over the following ten days it is gradually withdrawn into the abdominal cavity. But Aristotle's enduring contribution has less to do with what he observed and more to do with how he interpreted what he saw. He envisaged the unfertilised egg as a machine, ready to be switched on by copulation and the presence of the male's semen, and it was this line of thinking that led him to believe (incorrectly, and in contradiction to Hippocrates) that the white of the egg produced the chick and the yolk its nutrient. Aristotle was also obsessed with 'final causes' – the *purpose* of things. In a curiously unhelpful form of circular reasoning he claimed that the final cause of embryonic development was to produce an adult fowl. No explanation at all.

Many features of a bird's biology, including its supremely lightweight skeleton and the laying of hard-shelled eggs rather than giving birth to live young, are considered adaptations for flight. J. D. Meyer (1748–56) produced a series of remarkable illustrations showing both the whole bird, here a green woodpecker, and its skeleton.

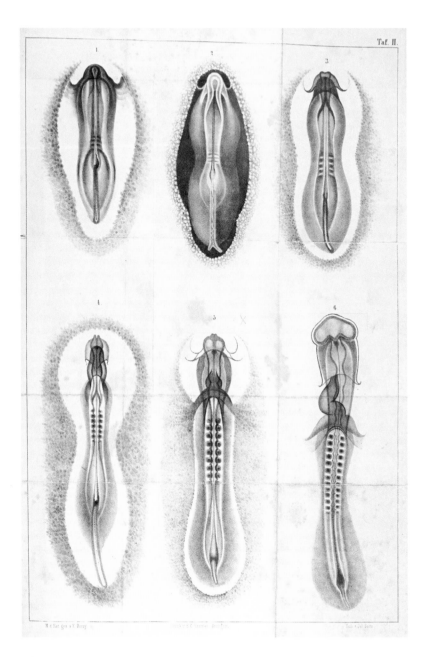

Birds' eggs provide a superb system for observing embryo development. These images show the development of what is called the primitive streak – the early stages of embryo development, with the head uppermost in each image. The top line shows (from left to right) the embryo after 21, 22 and 23 hours; the bottom row, after 25, 27 and 36 hours. (From Dursy, 1866)

The first truly fresh insight on the chick embryo was provided in the 1570s by a Dutchman, Volcher Coiter, who studied under a succession of Renaissance scholars, including Gabriele Falloppio, Guillaume Rondelet and Ulisse Aldrovandi. Inspired by Hippocratic advice and Aldrovandi's encouragement, Coiter was the first to make the much-needed day-by-day examination of the chick's development, accurately describing the structure of the germinal disc, and the egg tooth that appears on the twelfth day of development and subsequently helps the chick break out of the shell. Coiter's description of the developing chick is unique in Renaissance biology for being almost totally free from 'scholastic bias', that is, the influence of Aristotle's ancient authority; instead Coiter simply described what he saw.[27]

Aldrovandi also made a detailed day-by-day description of the developing chick, but was so convinced that what Aristotle said was true that his description is virtually useless. Then we come to Fabricius, who provides another error-strewn account of a chick's development. He thought each ovum was attached to the ovary by a little stalk and that, on being released, this left on the yolk a small white scar (the germinal disc, no less!) that played no further part in development. The embryo's starting point, he said, was the chalaza at the blunt end of the egg, and its subsequent development was like building a house: the framework (the skeleton) was established first and the heart, liver and lungs added later (and all at the same time).[28] Extraordinary! Fabricius's account includes some beautiful and extremely accurate illustrations of the chick's development, so how could he draw one thing and write another? One possibility is that Fabricius described what he believed, rather than what he saw. The other is that someone else later added the drawings to his account (which wasn't published until 1621 and after his death). Either way, Fabricius's written description of the developing chick was poor, and coloured by a set of beliefs, in defiance of what he must actually have seen with his own eyes. Self-delusion is the curse of science. Fortunately William Harvey was highly critical of Fabricius and rectified his many errors.

Watching the miracle of life unfold within the hen's egg in the early 1600s, Harvey noted how on the fourth day of incubation the heart of the embryo chick was just visible. He also observed the heart's sensitivity to cold and how, as the egg cooled, it gradually ceased to beat: 'I laid my warm finger upon it,' he wrote, 'and after the space of twenty pulses of my own artery, lo! the tiny heart awakened again and aroused itself, and, as if brought back from the threshold of the grave, resumed its former dance.'[29]

Incubating batches of eggs under broody hens, Harvey – like Coiter – systematically cracked them open on successive days to chart the course of the chick's development. His careful description was the best so far and he was dismayed by the errors of his predecessors, including Aristotle, who he decided had probably never looked at developing eggs himself, but must instead have relied on what others told him. How else could Aristotle make the monstrous blunder of saying that the white, or albumen, was 'the material constituting the chick'?[30] Also, what possessed Aldrovandi to ignore the excellent descriptions by his student Volcher Coiter in favour of his own antiquated interpretation? And how could Harvey's own master, the great Fabricius, fail to describe what he had obviously seen? The ancients' hold on progress was firm indeed and for the likes of Aldrovandi and Fabricius breaking free from Aristotle was extraordinarily difficult. By starting to sever that umbilical link with the past, Harvey became a forerunner of the scientific revolution.

A major issue for all embryologists in the 1600s was whether new individuals were preformed inside the sperm or ova and merely had to grow in size, or were created anew from the material inside the egg, in a process referred to as 'epigenesis'. Since there was no evidence either way, the debate between the preformationists (those that assumed the embryo was preformed and simply expanded in size) and the epigenecists raged for years. Even the preformationists were divided; the *ovists* thought that eggs contained preformed little chicks, while the *spermists* thought the same was true of sperm. The spermists were greatly encouraged in 1694 by Nicolas Hartsoeker's image of a little man (a homunculus) inside the head of a human sperm,[31] but there was also concern that if each of the millions of sperm within a human ejaculate contained a little person, God was being dreadfully wasteful.

Because he could not bring himself to believe that eggs contained miniature, preformed chickens, Harvey favoured the idea of epigenesis but was not really sure. The crux of the problem was this: on the one hand, epigenesis meant that a new organism was somehow created from 'nothing', and, on the other, preformation meant that within each egg or sperm was a miniature adult (and within that another and so on, like so many Russian dolls). Both views seemed implausible and it wasn't until 1759 that the issue was resolved, in favour of epigenesis. Some extraordinarily detailed observations of chicken embryos by Caspar Wolff in that year confirmed that there were no organs present in the very early stages of development and that when organs did start to appear they were not – crucially – in their final form.[32] If organs such as the heart and limbs could appear by epigenesis, it was no great step to imagine an entire individual developing in this way. But epigenesis was only the beginning; the daunting question of *how* new structures or a new individual is created, remained.

The chick was the perfect study system to see the day-to-day changes in a growing embryo, but it was less useful for seeing *how* those changes occurred. The eggs of other animals, notably frogs, starfish and sea urchins, were much more tractable in this respect, eventually revealing the processes of cell division and growth to nineteenth-century researchers. In the two millennia between Hippocrates and Harvey researchers did little more than provide an increasingly detailed picture of the successive changes that took place in a hen's egg between laying and hatching. There was no interpretation and no clue as to how these miraculous changes actually occurred. The descriptive tradition continued long after Harvey, and indeed was given new life in the eighteenth and nineteenth centuries, particularly in Germany, by the development of more sophisticated microscopes, and again in the 1940s by the electron microscope with its extraordinary magnification.[33]

The key to unlocking the secrets of development was experimentation: disrupting normal development and seeing what happened when things went wrong. Only in this way could researchers finally start to piece together the much-needed clues to normal development. By transplanting tiny bits of tissue from one part of the embryo to another, and by seeing whether, or how, they subsequently developed, researchers began to realise – contrary to anything they could have imagined – that clusters of cells within the embryo must be communicating with each other, with certain groups of cells dictating to others how they should develop. By labelling living cells with different-coloured dyes embryologists were able to see the flowing, pushing and bulging that went on among groups of cells as the chick began to take shape. This remarkable process, captured on film in the 1920s using time-lapse photography, provided stunning evidence for the extraordinary dynamic processes that occur as an embryo develops.[34]

The story of the chick's remarkable development was beautifully summarised by the Russian émigré Alexis Romanoff in the late 1940s. Born in St Petersburg in 1892, Romanoff worked first as a school-teacher and portrait painter, serving as a military engineer during the First World War, where he was on the side of the White Russians when the civil war broke out in 1917. When the Reds gained control in 1920 he borrowed civilian clothing, threw away his uniform and melted into obscurity, moving first to China and eventually to New York, where he turned up in 1921 with little more than his wits and a couple of references.

Entering Cornell University in 1923 at the age of thirty, Romanoff gained a B.Sc. in 1925 and his Ph.D. in 1928. As he put it, he 'simply fell in love with egg. You know why? Egg is wonderful creation.' It was while studying for his Ph.D. that he decided to write the definitive account of the egg's biology.

Twenty years in the making, *The Avian Egg* was a *tour de force*. The manuscript and 435 beautiful illustrations, all of which he had prepared himself, filled two suitcases. In 1947 Romanoff travelled from Cornell to New York City and presented himself and the suitcases to the publisher John Wiley. Impressed, but horrified at the prospect of such an enormous book, Wiley suggested he cut it by half. Romanoff

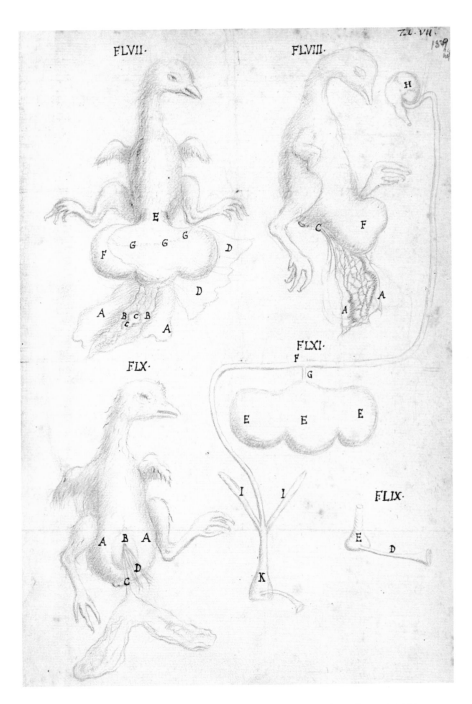

Another of Malpighi's accurate drawings of a chick embryo from the mid-1600s — here close to hatching, with the yolk sac clearly attached to the abdomen.

firmly but politely refused, explaining how he and his wife – his co-author – had spared nothing for the book. Luckily for science, Wiley relented and the 918-page book was published in 1949, to great acclaim. Romanoff even featured in *The New Yorker* and, when asked by a reporter how he had been able to produce such a comprehensive account, he said: 'I enjoy to beat other people by working harder. Others ask how can I do book like this even in twenty years. We work all the time – day, evening, weekend. Otherwise book would take twice more time …' They rarely took a holiday; had no children and no other distractions from their studies of eggs. *The Avian Egg* became the avian embryologists' bible.[35]

Ironically, by the time their second volume, *The Avian Embryo*, was published in 1960, the study of experimental embryology had lost its momentum and was slowly grinding to a halt. By the 1970s, when embryology was reinvigorated by the discovery of DNA technology, the chick had been eclipsed by much better model systems, including a tiny worm, the fruit fly and the mouse.

For a decade or so it looked as though the chick embryo had run its course. And then in 2004, fuelled by the publication of the chicken genome, phoenix-like the chick reasserted itself as a fabulous study organism. Little worms and fruit flies can tell us only so much about how human embryos develop – the chick on the other hand proved a perfect model system for revealing how a single cell (the ovum) is transformed into a warm-blooded vertebrate comprising millions of different types of cells, all containing the same genes.[36]

By failing to recognise that the microscope might be the key to unlocking the mystery of reproduction, William Harvey was left floundering and frustrated. Even after Leeuwenhoek made public his discovery that semen contained 'animalcules' and proposed that embryos were started by the combination of a single sperm and an egg, Ray couldn't bring himself to accept it. For once physico-theology let him down. The all-wise creator simply would not have been so profligate as to produce millions of sperm in semen if only one was needed to fertilise an egg. There seemed precious little wisdom in that and, because it didn't add up, Ray chose to ignore it.

As the quality and design of microscopes continued to improve throughout the eighteenth and nineteenth centuries, it became increasingly clear that semen of all animals contained sperm – millions of them – and it became harder to ignore the possibility that they played a crucial role in fertilisation. The problem was finally resolved in 1875 when the German biologist Oskar Hertwig, working with the beautifully transparent eggs of sea urchins, was able to witness a sperm penetrating an egg and fuse with the female nucleus inside it.[37]

It took rather longer for anyone to witness fertilisation in birds – not least because, unlike those of sea urchins, the eggs of birds are far from transparent. The huge amount of yolk in birds' eggs makes it particularly difficult to see fertilisation taking place. Success finally came – albeit indirectly – through the efforts of Charles Otis Whitman, one of the United States' most influential nineteenth-century zoologists, but now virtually forgotten. Whitman trained in Leipzig in the 1870s under the great German zoologist Rudolf Leuckart whose research on fertilisation inspired a passion for reproduction in his students. It wasn't until the 1890s, however, when Whitman was made professor of zoology at the University of Chicago and in his fifties, that he began to study birds. There was no funding for research and Whitman kept his birds – several hundred of them – at home and at his own expense. His results were astounding, something he attributed to the fact that he lived with his birds 'day and night year in and year out', a devotion sadly lacking in most modern ornithologists. For Whitman the pigeon was the perfect study species and his grand plan was to use it to bring together three different areas of biology: heredity, behaviour and development.[38] Put another way, Whitman wanted to discover everything there was to know about pigeons and to achieve this he divided up the responsibility among his brood of outstanding research students, among whom Eugene Harper was allocated the task of fertilisation.

Pigeons were ideal for this in one respect because, in contrast to most other birds, the timing of the laying of their two-egg clutch is so predictable. Since Aristotle's day at least it was known that a pigeon's first egg is typically laid in the late afternoon and the second one early in the afternoon two days later. By dissecting birds at different times Harper was able to see that the second ovum was released from the ovary just a few hours after the first egg had been laid. By focusing on the second ovum, removing the germinal disc and examining it under the microscope, Harper could see if fertilisation had occurred. Finding that the second ovum was fertilised almost before it was fully released from the ovary, Harper was the first to demonstrate that fertilisation in birds occurred in the uppermost region of the oviduct:

> The egg is clasped by the funnel-like mouth of the oviduct, which at this time has been observed to display active peristaltic [rhythmic] contractions, as if in the act of swallowing the egg ... Hence the entrance of spermatozoa must take place as soon as the germinal disc is exposed by rupture of the follicular wall.[39]

To test his idea, Harper took an *un*fertilised ovum from the ovary, carefully removed its germinal disc – and placed it in a dish with some fresh pigeon sperm. By ingeniously separating the germinal disc – which Malpighi had earlier shown to contain the female nucleus – from the obscuring yolk, Harper was, for the first time, able to witness fertilisation *in vitro* – literally, in glass. Success at last.

But Harper also discovered another extraordinary aspect of fertilisation in birds. Contrary to what happens in sea urchins and other organisms in which fertilisation had by then been observed, a hen's ovum is typically penetrated not by a single sperm, but by several. Multiple penetration may be associated with the fact that birds' yolk-filled ova are relatively huge and the germinal disc represents such a tiny target that several sperm are required to ensure that one of them reaches its target. Harper confirmed that, once inside the ovum, only a single sperm fuses with the female genetic material, just as it does in all other animals.

Harper's wonderful observations of pigeon fertilisation should have been the cause of great zoological celebration. But for some reason his results were virtually ignored. Even more remarkably, it was another thirty years before the same observations were made for the domestic fowl. When Marlow Olsen of the United States Department of Agriculture finally described the process of fertilisation in the fowl, he ignored Harper's pioneering studies;[40] possibly some kind of intellectual barrier existed between the poultry and pigeon biologists, or perhaps poultry biologists couldn't imagine fertilisation occurring in the same manner in the two types of bird. Olsen's research confirmed Harper's finding that each ovum is fertilised separately, but because chickens lay a succession of eggs at daily intervals, each ovum is fertilised about twenty-four hours before being laid.

Resolving the process of fertilisation was one thing; working out how birds managed to produce not just one but a succession of fertile eggs for days or weeks after the last copulation was more of a challenge.

In the early 1900s poultry biologists were still trying to decide between Fabricius's two explanations – simultaneous or sequential fertilisation – for the hen's protracted fertility. If they had only read Harper's account, and made the small ornithological step from pigeon to fowl, they would have resolved the problem at a stroke. Instead, they spent decades trying to distinguish between Fabricius's two ideas.

One particular experiment conducted during the 1920s seemed to confirm that a hen's ova were fertilised all at the same time. As ingenious as it was misleading, this study involved flushing the oviduct of recently inseminated hens with a spermicidal solution that would destroy the sperm. If the hens continued to lay fertile eggs after this the ova *must* be fertilised simultaneously. And that is exactly what happened; the females *did* continue to lay fertile eggs.[41] When poultry researchers later confirmed Harper's observation on the timing of fertilisation, they realised that simultaneous fertilisation was a physical impossibility (because for the sperm to gain access, the ovum has to be released from the ovary membranes), and only then began to wonder

Overleaf: *Considered an economic pest, the North American red-winged blackbird defied attempts to control its numbers in the 1970s by vasectomising the males because the females were promiscuous.* (From Catesby, 1731–43)

whether there might be an alternative explanation for the spermicidal results. Perhaps the sperm were hiding in crevices, as Fabricius suggested, and were thus protected from the spermicide.[42]

And a quarter of a century later, such crevices were found to exist! Govert van Drimmelen, a South African veterinary scientist, discovered what he called 'sperm nests' in the uppermost part of the oviduct (in the infundibulum).[43] At long last the great puzzle of Aristotle, Fabricius, Harvey and others seemed to be resolved: the infundibulum was where female birds stored their sperm, allowing them to produce a succession of fertile eggs long after the last copulation.

But it wasn't true. In the 1950s another poultry researcher, Peter Lake, working in Edinburgh, discovered another kind of sperm crevice, this time much further down the oviduct at the junction of the vagina and uterus. He noticed clusters of sperm packed inside tiny tubules, like sardines in a tin, except that all their heads were facing the same way. Preoccupied by other things, Lake merely noted what he had seen and carried on. But later, during a visit to the University of California at Davis in 1960, he happened to mention his sperm-tubules to a group of researchers studying fertility in chickens and turkeys. Recognising that the sperm in these tubules might provide the key to understanding the long duration of fertility, Wanda Bobr, a Polish Ph.D. student, immediately began to investigate and within a year had demonstrated convincingly that the sperm storage tubules were indeed the hen's main site of sperm storage.[44]

Van Drimmelen's 'sperm nests' in the infundibulum were a red herring. They were an artefact of artificially inseminating large numbers of sperm directly into the upper region of the hen's oviduct. Of course, since fertilisation takes place in the infundibulum *some* sperm must occur there following natural insemination, but this is not the main storage site for sperm.

The processes of sperm storage and fertilisation determined initially for the domestic pigeon and domestic fowl are now known to be virtually identical in all birds. It is easy with hindsight to say 'of course', but in the mid–late twentieth century it was far from obvious that this would be the case. The existence of sperm storage tubules in wild songbirds (passerines) was discovered only in the 1970s when the American biologist Olin Bray and colleagues were checking whether vasectomy could be used to control the numbers of red-winged blackbirds, an agricultural pest. To Bray's surprise (and disappointment), many of the female blackbirds paired to vasectomised males continued to produce fertile eggs – not because the vasectomy failed, but, contrary to all expectations, female red-winged blackbirds turned out to be highly promiscuous and appeared to copulate with males other than their partner. In Bray's words: '… examination of reproductive tracts showed that most birds … had sperm stored in the uterovaginal glands'. He provided no further details and did not illustrate the glands in question, but his novel findings subsequently got a mention in the scientific journal *Nature*, ensuring wide publicity for both the study but also the promiscuity of female blackbirds.[45]

The next step in the story of sperm storage in wild birds involved a young research student, Scott Hatch, who in 1979 started a Ph.D. at the University of California at Berkeley on the breeding biology of fulmars breeding in Alaska. The fact that there appeared to be a month-long delay between copulation and egg-laying made this species particularly intriguing. In 1981 Hatch was invited to give a seminar in the nearby Department of Avian Sciences – the same department that Peter Lake had visited in the 1960s – and after his talk was encouraged to talk to Frank Ogasawara about sperm storage in the fulmar. Hatch told me:

> To poultry scientists like Frank Ogasawara, the phenomenon of avian sperm storage was more or less old news, having been known for years and investigated in some detail in chickens and turkeys. Naturalists, it seemed, had not picked up on the possible significance of this for birds generally, and poultry scientists mostly talked to other poultry scientists about poultry science. At that first meeting, Frank and I went out to

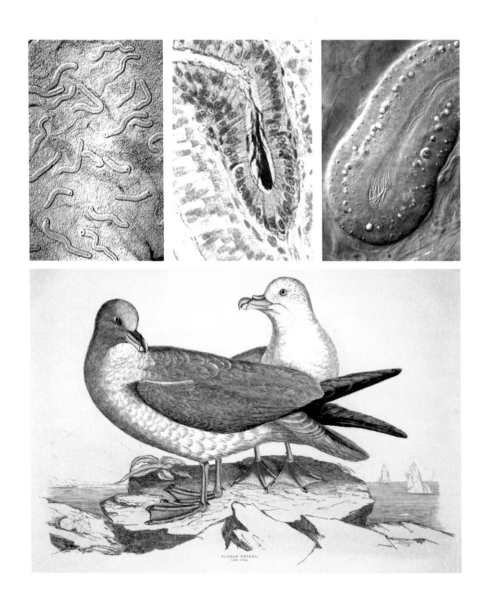

Sperm-storage tubules (from left to right): several entire tubules (average length 0.3 mm) of a Japanese quail; a single fulmar petrel tubule containing a dark mass of sperm (from Scott Hatch's original study); and detail of a single turkey tubule containing sperm. A pair of fulmar petrels (from Selby 1825–41).

the Department's holding pens, where he unceremoniously dispatched one of his chickens and proceeded to demonstrate how to locate the so-called 'UV' region* in the gross anatomy.[46]

It is now known that the females of all birds release their ova sequentially from the ovary, each one twenty-four to forty-eight hours before it is laid. Since an ovum can be fertilised only during the fifteen-minute period while it is in the infundibulum, sperm storage – which ensures a constant supply of sperm delivered to the infundibulum – avoids the inconvenience (or impracticality) of having to copulate at a very specific time each day to fertilise each egg in turn.[47] Sperm storage is a ubiquitous feature of birds, but varies in duration from around a week in pigeons and doves to over a month in fulmars and other species where male and female spend long periods apart. A convenient arrangement.

The result of fertilisation is an embryo that develops rapidly inside the egg. It is a different world when the young bird hatches out of its protective shell. Depending on the species, chicks emerge either on their own two feet and with their eyes wide open, as with young chickens or ducklings, or, in the case of songbirds, naked, blind and virtually helpless. Regardless of what they are like when they hatch, the young bird relies on its behaviour to help it through to independence. In the next chapter we will look at how young birds acquire these crucial behaviours.

---

* Utero-vaginal region – the junction between the uterus and vagina.

The serious study of animal behaviour began only in the mid-twentieth century. Until then, the way in which young birds – like these twelve-day-old, recently fledged garden warblers – acquire the behaviours they need to see them through the rest of their life was a complete mystery. (Painting by Henrik Grönvold, from Howard's British Warblers, *1912)*

3

Preparation for Life

Instinct and Intelligence

Calling someone bird-brained means they are silly or stupid, but it also implies that most people think birds are generally not very smart. Nothing could be further from the truth, as the following incident, which occurred while I was living in California with some researchers studying acorn woodpeckers, demonstrates.

An intensely social, group-living bird, the acorn woodpecker is best known for its spectacular granaries of stored acorns. Groups comprise a pair of breeding birds and several male offspring from previous breeding seasons that stay at home to help their parents rear subsequent broods. Young females, however, leave home and have to join another group. It was this dispersal behaviour my house-mates were studying, keeping track of radio-tagged, colour-ringed females as they prepared to leave their natal group. One day the researchers came back unable to contain their excitement at what they had seen. A female they had been watching for several days had set off that morning, flying some ten kilometres (seven miles) directly to another woodpecker group that she apparently wanted to join. The group's resident males had other ideas and chased her away. The female flew home again and began interacting noisily with her brothers. Later, accompanied by her brothers the female set off, flying straight back to the new group, where the female's brothers attacked the resident males, and departed only once their sister had been accepted by the group!

Where does such sophisticated behaviour come from? Is it innate, or do acorn woodpeckers have to learn it? The question of whether bird behaviours derive from nature or nurture is a long-standing one and caused centuries of confusion. The notion that birds often behave in a silly or inappropriate way is usually a consequence of seeing their behaviours out of context, and one of the most striking examples of this is the behaviour of young chickens or geese immediately on hatching. Ever since people started keeping poultry it has been known that young birds will follow a human foster parent as though it was their mother.

During the 1930s Konrad Lorenz studied this behaviour in detail and used the term 'imprinting' to describe the attachment young geese showed to him, and his provocative paper on the topic was a landmark in the study of animal behaviour:

> To the uninitiated it is often surprising (even incredible) that a bird does not recognise conspecifics [members of its own species] purely 'instinctively' in all situations ... The behaviour appears so bizarre ... that any individual observer who encounters this phenomenon when hand-rearing young birds is at first prone to regard it as a pathological process, explained as a 'confinement psychosis' or the like.[1]

As Lorenz well knew, imprinting seemed pathological or silly only when it occurs in an inappropriate context, as in captivity where the first thing a gosling sees on hatching is a human. In nature, of course, the first thing a gosling sees is its mother, in which case its attachment might be interpreted as very clever, but it might also be seen as either 'instinctive' (innate) or learnt. Lorenz's pioneering research provided the first answers to these fundamental questions about how young birds and other animals acquire the behaviours that prepare them so beautifully for adult life.

Imprinting has been known for a very long time for Pliny describes instances of geese becoming emotionally attached to their owners:

Pioneer animal behaviourist and later Nobel Laureate Konrad Lorenz,
followed by an imprinted brood of goslings.

EIDER DUCK, MALE.

At Aegium [a goose] is said to have conceived a passion for a beautiful boy, a native of Olenos, and another for Glauce, a damsel … One might be tempted to think that these creatures have an appreciation of wisdom: for it is said, that one of them was the constant companion of the philosopher, Lacydes, and would never leave him, either in public or when at the bath, by night or day.[2]

In the seventh century St Cuthbert, Bishop of Holy Island, who became a hermit on the Farne Islands, in north-east England, had hand-reared eider ducks that followed him and ran to him in times of danger. The species was locally important because of its down and St Cuthbert's chronicler Reginald of Durham referred to the species in *c*.1165 as 'aves … Beati Cuthberti', or birds of the Blessed Cuthbert, and until the mid-1600s the bird was generally known as the Cuthbert duck.[3]

When in 1516 Sir Thomas More described the marvel of artificial incubation of chicken eggs by English farmers in his book *Utopia*, he noted that 'as soon as the chicks come out of the eggs, they follow the men and recognise them as if they were their mothers'.

Imprinting is now considered largely a learning process in young birds (it also occurs in other animals, including ourselves) that involves a very rapid attachment during a 'sensitive period' to either a mother figure and/or a future mating partner.[4] As this implies, there are two types of imprinting, one between a young individual and its parent, known as *filial imprinting*, and the other an expression of sexual preference referred to as *sexual imprinting*. The evolutionary significance of each is obvious. In the first case it provides a young animal with crucial information about its parents' identity, ensuring that it is properly cared for. In the second, imprinting ensures the appropriate choice of partner in later life.

Although the phenomenon of filial imprinting was well known (if not by name) to poultry-keepers, the first experimental studies to understand it were conducted only in the 1870s by Douglas Spalding. Born to working-class parents, Spalding educated himself

Previous pages: *Known originally as St Cuthbert's duck, the eider duck was so named because some imprinted individuals appeared to be under the spell of St Cuthbert, Bishop of Lindisfarne, who died in* AD 687. *This painting of a male eider is by Selby (1825–41).*

while employed as a labourer. He studied literature and philosophy at the University of Aberdeen in 1862, eventually qualifying as a lawyer. A meeting with the philosopher and social reformer John Stuart Mill in the late 1860s transformed Spalding into a dedicated and extraordinarily talented amateur scientist. To assess how it affected their tendency to follow him, Spalding placed little hoods on domestic fowl chicks immediately after hatching and then removed them at varying times. One group of chicks that wore hoods for four days, on having their hoods removed ran from Spalding in terror on seeing him for the first time. This was extraordinary, for as he commented:

> Whatever might have been the meaning of the marked change in their mental constitution – had they been unhooded on the previous day they would have run to me instead of from me – it could not have been the effect of experience; it must have resulted wholly from changes in their own organisation.[5]

Here was the first evidence of the existence of what we now call a 'sensitive period', meaning that if imprinting did not occur in the first three days after hatching, it did not occur at all. Spalding concluded that, on hatching, chicks had an instinct to follow, and would follow anything, but they also had an instinctive recognition of their mother's voice to ensure that they followed the right thing. He also showed that there was a 'sensitive period' for imprinting on the hen's voice: 'A chick that has not heard the call of the mother until eight or ten days old then hears it as if it heard it not.'[6]

Spalding's contribution to the study of animal behaviour – particularly to development, instinct and imprinting – was unparalleled at the time. In the studies just described, to ensure that the chicks had no visual experience of anything before his experiments Spalding incubated eggs himself by suspending them in a bag over a kettle emitting warm steam, and opened the eggs when the chick's eyes were still closed (a day or so before it would normally hatch), and then, carefully straightening the bird's neck, secured a translucent hood over its head with an elastic band. Spalding knew that although the hood prevented the chick from feeding, once the chick had hatched it could survive perfectly well for several days on its yolk reserves.[7]

What marks Spalding as exceptional is that he was the first to recognise that the *only* way to distinguish the effects of nature and nurture in behaviour was by means of experiments. His first scientific paper, on instinct, presented at the British Association for the Advancement of Science at Brighton in 1872, was widely praised. An extended version of the lecture was reprinted in a popular science magazine and attracted the attention of the philosopher George Henry Lewes, who wrote: 'Mr Spalding has not only proved himself an acute thinker, he has shown a rare ability in devising experiments, and we may fairly expect his researches will mark an epoch.' It certainly did, for Spalding's work not only signalled the beginning of the experimental study of animal behaviour, but it also initiated ferocious opposition from the emerging (and competing) field of psychology. In some ways Spalding brought this upon himself for he was highly critical of the psychologists' anecdotal approach and their reluctance to conduct experiments; he also questioned 'the very pillars of their philosophical edifices, consciousness and feeling'. Inevitably he was marginalised, and to such an extent that it eventually obscured his contribution as a scientist. He died of tuberculosis at the age of thirty-seven.[8]

The idea that sexual preferences may be set, or at least influenced from an early age, was recognised by Leonard Mascall in his 1581 book on poultry husbandry, when he reported that male ducks reared under domestic hens later exhibited a 'desire to tread the hennes'.[9] Subsequent observations by bird-breeders revealed such sexual imprinting to be widespread.[10] From the 1600s, as cage-birds became increasingly popular, it was fashionable to create hybrids between different finch species, especially between a finch and a canary. It was not always easy, but it soon became obvious that one particular trick increased the chances of success. This was to rear finch chicks in the nest of a canary, so that when the male finch matured it would more readily

Hybrids between the canary and goldfinch (top left), linnet (top right) and bullfinch (lower) were often created by fostering young finches under canaries, so that the finches imprinted on their foster parents and readily copulated with them when sexually mature. (From Robson and Lewer, 1911)

copulate with female canaries. Buffon's main informant about such things was a Father Bougot, who said that to create goldfinch 'mules' (that is, hybrids between a finch and a canary) one had to raise the goldfinch in the nest of the canary and keep other goldfinches away.[11] Similarly, when Charles Whitman (who was probably completely un-aware of that method of obtaining canary mules) reared various wild-pigeon species under domestic ring doves in the early 1900s, he found that as adults they preferred to breed with ring doves and, as a result, produced hybrids.[12]

Exactly the same kind of imprinting can also occur between birds and people. Many of the birds hand-reared by Lorenz later regarded him as a prospective partner – his jackdaws, for example, attempted to courtship-feed him by placing food in his ear.[13] It is not clear whether Lorenz's jackdaws or geese ever solicited copulation or attempted to copulate with him; if they did he never wrote about it. Australian ornithologist Richard Zann had no such reservations and described how a male zebra finch he had hand-reared in the 1960s attempted to copulate with, and ejaculated on to, his finger.[14] A male zebra finch that my daughter hand-raised routinely performed courtship song to her, but, possibly because it was blind from birth, never attempted to copulate. Falconers have known since at least the thirteenth century that young raptors can become sexually imprinted on their owners and sometimes attempt to copulate with them. Indeed, since the 1970s fal-coners have exploited this behaviour by arranging for birds to imprint on a hat, for example, so that as adults they would copulate with it and hence produce a semen sample that could then be used for artificial insemination and captive breeding.[15]

His experiences with jackdaws and geese led Lorenz to conclude that the main consequence of imprinting in nature was the determina-tion of adult sexual preferences – ensuring that individuals reproduce with their own kind when they mature. This is true, but how do in-dividuals avoid sexually imprinting on their parents or brothers and sisters? The answer was provided by Cambridge zoologist Pat Bateson, who discovered from studies of Japanese quail in the 1970s that young birds develop preferences for members of the opposite sex that are slightly different – but not *too* different – from individuals with which they had been reared; a kind of optimal out-breeding, ensuring that birds avoid mating with close relatives.[16]

Building on Spalding's ingenious studies Konrad Lorenz made the subject of imprinting his own and it was largely for this work that he (together with Niko Tinbergen and Karl von Frisch) was awarded the Nobel Prize in 1973. His early understanding of behaviour was based almost entirely on the close observation of animals he kept at his home-cum-research station at Altenberg, Austria, as a boy. Lorenz was a genius. His perception, insight and knowledge of the behaviour of the animals he kept were remarkable. Even so, Lorenz was reluctant to perform experiments to test his ideas, as Spalding had done. Relying on observations alone, it was perhaps inevitable that he would make mistakes.

Lorenz concluded, for example, that imprinting differed from other kinds of learning in that it did not require any reward or reinforcement; it occurred during a 'sensitive period' and was irreversible, so that once an animal had formed a social bond with one individual it could not do so with another. Lorenz was too dogmatic. Imprinting is not, it turns out, fundamentally different from other forms of learning and the reward or reinforcement the young animal obtains is the imprinting object itself – usually the animal's mother. In terms of sensitive periods, Lorenz's notion that if an individual obtained the 'wrong' information during the brief period in early life then its behaviour was irreversible was also wrong. Since Lorenz's time studies of a variety of birds and other animals have revealed quite a lot of flexibility in both filial and sexual imprinting.[17]

Lorenz's idea of sensitive periods, however, was basically correct. He and other early ethologists likened a young bird to a passenger travelling through time on a train whose windows are opaque. At a predetermined time a window opens briefly to display the landscape, and then closes again. Whatever the young bird sees during the short period when the window is open shapes its future. In reality there are different windows that open and close at different stages of the journey and remain open for varying periods of time. It later became clear that the windows – the sensitive periods – are controlled either by an internal clock (just as with migration and song acquisition, as we shall see) or as a result of experience. Not surprisingly, then, the periods during which filial and sexual imprinting occur usually differ and, as Spalding showed with his hooded chicks, filial imprinting typically takes place early in life, while sexual imprinting occurs later and this particular window remains open for much longer, sometimes right up to the time of mating.

Although not all species show the same following behaviour as Lorenz's goslings, the majority of birds – regardless of whether they are 'precocial', that is, hatch with their eyes open like geese and ducks, or 'altricial' and hatch helpless and naked with their eyes closed like finches and sparrows – exhibit sexual imprinting to a lesser or greater extent. There is one striking exception to this, however. For brood parasites like the European cuckoo and North American cowbird it would obviously be disastrous if their young imprinted *sexually* on their

The best known of all brood parasites, the European cuckoo exploits a range of host species and avoids sexually imprinting on them. (From Frisch, 1743–63)

foster parents. On the other hand young brood parasites must imprint on them *socially* so that they know which species to parasitise in later life. How they know who to copulate with, though, remains a mystery but it must be instinctive.

One of the earliest studies of imprinting in brood-parasitic birds was conducted on African whydah birds, which are extraordinary in that each species parasitises a single host species (an estrildid finch), unlike the European cuckoo, which utilises a range of host species. The paradise whydah, for example, parasitises only the Melba finch. The question is, how does it do it? How does the female whydah know which species to parasitise, and how does she know which male to mate with? The answer, at least in part, is that young whydahs of both sexes imprint on the song of their estrildid foster parent. When the male whydah matures it sings the estrildid song, which the female whydah uses both as a guide as to who to pair with, but also who to parasitise.

This still leaves the issue of which one they should mate with. Recent research on the North American cowbird, suggests that young parasites recognise their own species by means of a special 'password' – a signal that may be a call, or some other behaviour unique to each species, that stimulates the young brood parasite to learn other aspects of its true identity and thereby make sure it mates with the right species.[18]

During the 1930s the study of imprinting was a major focus of the entire field of animal behaviour. The broader issues, of course, were instinct and intelligence and the centuries-old question of man's place in nature. With none of the constraints imposed by Christianity, the ancient Greeks saw animals and humans as part of the same scheme, but the medieval Church insisted that humans were both distinct and superior. Beasts and flowers had been placed on earth solely for man's benefit, to be used and abused with impunity. From the mid-1500s the Church's grip began to loosen, partly a result of the increased awareness of animals resulting from the encyclopaedias of Gessner, Belon and Aldrovandi, and animals started to gain some moral respect – that is, until René Descartes began to espouse his views. A key figure in the scientific revolution – the search for reason and truth and the sweeping away of magic and folklore – that began in the early 1600s, Descartes found himself caught between a rock and a hard place: his Catholic faith (and fear of the Inquisition) on the one hand and science on the other. He compromised by claiming animals to be soulless automata whose behaviour was entirely instinctive. As anyone who had closely watched birds or other animals knew, this was ridiculous, and one of Descartes's most vehement opponents was John Ray:

> This opinion, I say, I can hardly digest … I should rather think animals to be endowed with a lower degree of reason, than that they are mere machines … Should it be true, that beasts are automata or machines, they could have no sense or perception of pleasure or pain, and consequently no cruelty could be exercised towards them; which is contrary to the doleful significations they make when beaten or tormented, and contrary to the common sense of mankind.[19]

Part of Ray's dismay was that Descartes's view seemed – paradoxically – to deny the existence of God. For Ray and other Protestants, natural history in all its sophistication was incontrovertible evidence for God, whose wisdom was manifest in the entire suite of behaviours exhibited by birds regardless of whether these were instinctive or sometimes appeared 'intelligent'. Some of the instinctive behaviours Ray described included the fact that when parent birds bring food for their chicks none of them is overlooked, even though (he said) they have no ability to count. God's wisdom was also evident in the courage of parents defending their young – a thing 'contrary to any notions of sense or instinct of self-preservation'. In addition Ray described what he considered to be strange instincts; the fact that when poultry and partridges utter an alarm call in response to a predator, their young hide themselves even though they may never previously have seen a predator or heard these alarm calls before. A long-standing question

1. La grande l'euve d'Angola, réduite.
2. La même l'euve, après la Mue, de grandeur naturelle.

The paradise whydah is another brood parasite, but exploits just one species, the Melba finch, as its host. The young parasites learn the host's song in order to identify who to avoid mating with, but who to parasitise later in life. (From Buffon, 1778)

Among the most intricate and exquisite of all nests, that of the long-tailed tit is a lichen-covered structure bound with spider silk and lined with hundreds of feathers. (From Hayes, 1825)

in the study of birds was whether nest-building was instinctive or something that young birds had to learn by watching older birds. Ray and Willughby were firmly on the side of instinct, pointing out that birds raised by hand and with no opportunity to watch older birds nest-build still produced a typical nest, providing clear evidence that nest construction is largely innate.[20]

Small birds have been kept as pets since the earliest times and much of their appeal has been the ability to perform special tricks. In Europe one of the best known was reported by Pliny and beautifully illustrated by the seventeenth-century German painter Abraham Mignon, and involved a goldfinch pulling up a tiny bucket of water or seed on a string in order to feed itself.

Another popular trick involved getting small birds to tell people's fortune. The species most often used in Japan in a particularly sophisticated performance is the varied tit. The fortune-telling routine mimics that of people visiting a shrine: the bird takes a coin from the trainer and places it inside a miniature collection box. It then hops up the steps towards the model shrine and rings a bell. Opening the shrine door, the bird pops inside, picks up a folded piece of paper bearing a message and returns along the causeway. It tears the red paper wrapper from the message and turns the paper in its bill as many times as it is told. Finally, the bird flies to the trainer with the fortune-telling paper in its beak and gives it to the trainer, who rewards the bird and then passes the message to his client.[21]

Our fascination with the ability of birds and other animals to perform tricks panders to our obsession with humanity's place in the world. It is curious how for much of our history we have been concerned with how distinct we are from animals and how much cleverer we are than them. But we love clever animals! Performing animals gain our admiration and respect, and in a way their 'cleverness' draws us closer, allowing us to anthropomorphise and become sentimental about them.

Bird species differ in how easily they learn, but we now interpret the differences in temperament and propensity to learn (and, as a result, to perform tricks) in terms of natural selection. The gold-finch's string-pulling trick, which was once regarded as a clear sign of its 'intelligence', is little more than an extension of its natural way of dealing with grass stems, pulling them through its feet in order to reach the seeds at the tip. In a similar way, the varied tit's fortune-telling performance is very similar to its behaviour in the wild, hiding and recovering seeds and tearing off papery seed casings to reach a food reward. Tits also have sophisticated cognitive ability, providing them with a good spatial memory and allowing them to recall where they have hidden seeds.[22]

One reason our predecessors found it so difficult to make sense of tricks like these was because the terminology they used to describe them was so horribly muddled that it almost guaranteed confusion. The terms 'innate' or 'instinctive' (literally, driven from within) were reasonably clear, and beautifully illustrated by the ability of a hand-reared bird to build a perfect nest. It was terms like 'intelligence' and 'reason' that caused confusion (they still do), epitomised in Pliny's statement about the *instinctive cleverness* displayed by nest-building birds:

> The swallow builds its nest of mud, and strengthens it with straws. If mud happens to fail, it soaks itself with a quantity of water, which it then shakes from off its feathers into the dust. It lines the inside of the nest with soft feathers and wool, to keep the eggs warm, and in order that the nest may not be hard and rough to its young when hatched.[23]

In the past when birds or other animals did something sensible, such as build an intricate nest or perform a clever trick, their actions were often regarded as intelligent or involving reason. Today we would describe nest-building as adaptive without necessarily implying anything about whether it was innate or otherwise. The crux of the issue was whether learning was 'intelligent' or evidence of 'reason'. At one time a continuum was thought to exist, with instincts (assumed to be genetically determined) at one end and learning, reason and intelligence

*The Japanese varied tit has been trained to perform an elaborate fortune-telling trick, exploiting its natural tendencies to hide and recover seeds.
(From Temminck and Schlegel, 1845–5)*

at the other: nature and nurture, respectively. In Darwin's words: 'Cuvier maintained that instinct and intelligence stand in an inverse ratio to each other.'[24] The truth, as Darwin recognised, is far more complex. Where a particular species lay on this continuum was one way animals were evaluated; the closer to the reason end, and hence to humans, the more worthy they were deemed to be.

Of all the tricks birds perform, none is more impressive than the ability to speak. Here is what Ray and Willughby had to say:

> Many birds are very ingenious and docile, as may appear from that they are so easily taught to imitate man's voice, and speak articulately: which no quadruped (for ought I have seen or heard) could ever be brought to; though their organs seem to be much fitter for that purpose, as being much more conformable to mans.[25]

They mention several accounts of talking parrots in their encyclopaedia. One of these came from the French botanist and explorer Charles de Lécluse, who told of a bird that when instructed *Riez, perroquet, riez* (Laugh, parrot, laugh) would do so on command and then say, 'O great fool, who makes me laugh', to everyone's delight. Another story first appeared in Conrad Gessner's encyclopaedia of 1555, and was designed to illustrate the uncanny ability of some parrots to utter the right thing at the right time – a trait often perceived as intelligence. A parrot belonging to Henry VIII fell out of a Westminster Palace window into the Thames. The bird, conveniently recalling a phrase it had heard previously, cried out "A boat, a boat for twenty pound". As a result the parrot was rescued by a passing boatman and restored to the king.[26] Apocryphal perhaps, but the ability of parrots to link certain objects or events with particular phrases has continued to fascinate us. Writing in 1790, after recounting a more believable example, William Smellie, naturalist, philosopher and translator of Buffon's works, commented:

> In this, and many other similar cases, the objects and the sounds are evidently connected in the mind of the animal. How far these associa-

tions might be carried by a patient and persevering education, it is difficult to determine. In this manner, however, parrots might be taught a considerable vocabulary of considerable nouns, or the proper names of common objects. But his intellect, it is more probable, would never reach the use of the verb, and other parts of speech.[27]

Two centuries later a scientist provided one particular parrot with exactly the patient and persevering education Smellie was thinking of, revealing that the parrots' cognitive powers were far greater than Gessner, Willughby, Ray, or anyone else for that matter, ever anticipated.

John Ray's view that birds and other animals were *not* fundamentally distinct from man gradually gained ground. This was due largely to the writings of ornithologists like Baron von Pernau, Johann Zorn and Johann Leonard Frisch in the early 1700s, who, inspired by *The Wisdom of God*, observed, recorded and attempted to interpret the behaviour of birds. There was also progress on a more philosophical front when the brilliant philosopher David Hume attacked Descartes's ideas: 'No truth appears to me more evident, than that beasts are endowed with thought and reason as well as men.' And as the science historian Philip Gray, writing in the 1960s, observed:

> It is a curious fact that in lands and among people where animals are treated as mere objects of utilisation, which was seldom the case in Hume's England, no great discoveries are ever made about animal behaviour and very few about human behaviour.[28]

The confusion that reigned over instinct and learning is epitomised by the philosopher Abbé de Condillac, whose book *Traité des Animaux* contains the statement that instinctive behaviour consists of 'acquired ability, developed from experience'. What he is saying here is that instincts are learnt – confirming that much of the misunderstanding about animal behaviour stemmed from a lack of definitions, and, in Abbé de Condillac's case, utter confusion over instinct and learning.[29]

William Smellie devised a sensible classification of instincts in his *Philosophy of Natural History* in the late 1700s, which ranged from 'pure' instincts (such as voiding faeces and urine, and sneezing); through those that can 'accommodate themselves to peculiar circumstances and situations' (e.g. birds building their nests at the ends of branches in areas inhabited by monkeys, but not elsewhere), and lastly 'instincts which are improvable by experience and observation' by which he means learning. In all but name Smellie distinguishes between instinct and learning, but he also implicitly recognises that learning may have an inherited component. Sadly, it is not obvious that at the time anyone appreciated Smellie's insights.[30]

In a desperate attempt to resolve once and for all whether men and beasts were distinct, Julien Offray de la Mettrie, in the mid-eighteenth century, decided to train animals to see just how much they could learn. His results showed that animals had only a limited capacity, confirming to him at least the fundamental distinction between humans and non-humans.[31] But one person interpreted this in a different way. Charles Georges Leroy, Louis XV's master of the hunt, pointed out that Mettrie's results were exactly what one would predict: animals learnt only enough for their particular lifestyle and it was unreasonable to expect them to aspire to anything greater – a remarkably modern, evolutionary kind of interpretation.

Charles Georges Leroy was an extraordinary combination of naturalist, thinker and writer. Moving effortlessly between *la chasse* (the hunt) and Parisian intellectual circles, Leroy knew and was cited by Buffon. Above all he knew animals and his tremendous insight into animal behaviour was based largely on observations made during hunting. His book *The Intelligence and Perfectibility of Animals From a Philosophic Point of View* appeared in an English translation in 1870, but his ideas – some of them extremely radical – were published originally under a pseudonym, 'the Naturalist of Nuremberg', in the 1760s. The book is written as a series of letters to 'a lady, comtesse d'Angiviller', with whom he was 'intimate' (or so it says in the preface of the English edition):

> It is some time, Madame, since you asked me for the letters of the Naturalist of Nuremberg on animals and men. I have had much difficulty in meeting with this little work, now becoming rare; but at last have the pleasure of sending it you. You will have it that I, too, have made some observations and reflections on the nature of animals, and you desire me to communicate them to you. Whatever you wish must be done … so I add to the letters of the Naturalist some few of my own …[32]

In his history of animal behaviour published in 1979 W. H. (Bill) Thorpe declared Leroy to be an important figure, not least because he anticipated the idea of the 'ethogram' – a summary of a species' behavioural repertoire.[33] The ethogram was a fundamental concept for the pioneers of animal behaviour, or ethologists as they called themselves. Leroy expressed it slightly differently, but his meaning is clear: 'I should like the biography of every animal', he said. Significantly, he recognised also that instinct, which he referred to as 'a kind of hereditary aptness or inclination to certain acts', together with learning were important determinants of an animal's behaviour. He totally rejected Descartes's notion of automata – which still wasn't dead: 'When you have studied a large number of individuals, chosen from different species … it seems impossible not to reject entirely the idea of automatism – of their being mere machines.' And if animals were automata, he said, they would behave in exactly the same way in the same situation. They do not, but people believe they do because, unlike him, they hadn't watched them closely enough: '[T]his uniformity is not so great as at first sight it appears; that we judge wrongly of it, from want of sufficient observation; and that possibly we have not the means of forming a correct judgement of it'.[34]

The term 'perfectibility' in the title of Leroy's book refers to the ability of animals to modify their behaviour through experience, and 'improve' themselves, much as people might. He also has a subtle

appreciation that an animal's behaviour is appropriate for its particular lifestyle, in other words that behaviour is adaptive. Darwin read
Leroy's book but makes little comment on it, which is surprising since
Leroy anticipates several Darwinian concepts, including adaptation
and the idea that females choose their partners, a central tenet of Darwin's idea of sexual selection. Leroy also foresaw some other important
concepts, as is apparent in this remarkable passage on the notion of
selfish genes:

> In those species in which the solicitude of the parents is closely con-
> centred upon the interests of the family, we see no affection for the spe-
> cies in general; rather, a decided hatred for those of their fellows who
> are not of the family ... The partridge, who is most careful and active
> for the welfare of her own brood, pursues and kills without pity all those
> who do not belong to her ...[35]

Another reason Bill Thorpe liked Leroy was because he appreciated
the value of detailed observation, claiming that:

> None but a sportsman [i.e. a hunter] can fully appreciate the intelli-
> gence of animals. To know them thoroughly, you must have associated
> with them; and most philosophers fail in this point ... Sportsmen, who
> observe because they have every opportunity, have rarely the time or
> the capacity to draw inferences, and philosophers, who reason without
> end, have rarely the opportunity of observing.[36]

Leroy's most enduring legacy is the notion that a lot of animal behaviour is what he called 'the product of reason', that is, involved
something approaching consciousness, and it may be because of this
that Darwin chose to say little about his ideas. In a rather bland review
of Leroy's book, Alfred Russel Wallace described Leroy as a 'loving
student of nature who observed and thought for himself, and who, in
many of his conceptions, was far in advance of the great philosophers
of the last century, among whom he lived'.[37] Erwin Stresemann was less

positive and berates Leroy for encouraging an unnecessarily romantic view of animal behaviour that later resulted in books such as Scheit-lin's *Science of Animal Souls* of 1840 which claimed that 'sympathy, compassion, love and hate, gratitude, vanity, reverence, conceit and pride' are all manifest in animal behaviour. Reaction to Descartes's view of animals was causing the pendulum to swing too far in the op-posite direction. As Stresemann noted despairingly, these ideas were 'swallowed by a romantic generation, who saw themselves in the new role of animal protectors'.[38]

The publication of Darwin's *Origin* in 1859 marked the end of the sharp divide between humans and non-humans. The confusion over instinct and intelligence remained, but Darwin's ideas breathed new life (and not a little controversy) into the study of animal be-haviour that Ray had initiated a century and half earlier. Darwin's view of instinct, as we might expect, was restrained. He recognised that many instincts, including self-preservation, maternal care, sexual behaviour and the suckling behaviour of young mammals, are com-mon to both non-humans and humans, but he was also in no doubt that many 'lower animals, like man, manifestly feel pleasure and pain, happiness and misery'.[39] Darwin, however, managed to sidestep the anthropomorphism of Scheitlin and others, neatly avoiding the instinct–intelligence debate and pointing out that the notion of an inverse ratio between instinct and intelligence was simplistic.[40]

Perhaps inevitably, Darwin's demonstration that we are all animals encouraged a romanticised view of nature, and birds in particular. In England George Romanes accumulated evidence in the form of end-less, fatuous anecdotes of the parallel intelligence of man and animals. Alfred E. Brehm did much the same in Germany. His father was the well-known and much respected academic ornithologist Christian Ludwig Brehm, active during the so-called golden age of field orni-thology between 1820 and 1850.[41] A country pastor, Brehm senior – like Ray – saw the remarkable relationships between form and func-tion in birds as part of God's wisdom: '[T]he more religious our hearts are, so much more the veil is lifted that conceals God's working from

The various display postures performed by males – like the blackcap shown here – during the breeding season towards females or rival males led early ornithologists to ask whether such behaviours were innate or learnt, and whether birds experienced emotions in the same way as humans. (From Howard's British Warblers, *1909)*

our dim eyes.' It is easy to dismiss ornithology informed or motivated by religion, but C. L. Brehm's studies of anatomical adaptations and the structure of birds were a major contribution to ornithology. His son Alfred had a rather different view of the world. An admirer of Scheitlin's anthropomorphism and seduced by the lure of his own popularity, Brehm junior promoted the idea that birds possessed character, reason and intellect and were capable of experiencing emotions. By saying that birds could *learn* to do *instinctive* things such as nest-building he perpetuated the ongoing muddle over instinct and learning. Brehm junior's contribution to the understanding of bird behaviour was mixed at best. To his credit he encouraged a sympathetic interest in birds, but he did so by capitalising on the popularity that natural philosophy enjoyed during the first half of the nineteenth century. But Brehm junior was also a contradiction: on the one hand he promoted Darwin's ideas, while on the other his romanticism was completely at odds with the development of scientific ornithology. A contemporary German psychologist referred to Brehm's work as 'vulgar psychology' and one can almost hear Erwin Stresemann's frustrated sigh as he wrote about Alfred Brehm's careless exuberance.[42]

Alfred Brehm's unlikely combination of Darwinism and anthropomorphism brought him into direct conflict with Bernard Altum, president of the Deutsche Ornithologen-Gesellschaft (the society of German ornithologists). Altum had trained as a clergyman (eventually becoming curate of Munster Cathedral), and not surprisingly was vehemently anti-Darwinian. But he also opposed Brehm's ornithological romanticism: 'The animal does not think, does not reflect, does not establish aims for itself and if it nevertheless behaves purposively, then someone else must have thought for it.' In other words, birds have no morals or reason and anything sensible that they appear to do must be guided by God. Despite his religiosity Altum was a good researcher. His antipathy towards Darwin was ineffectual, but his diatribe against Brehm's romanticism was instrumental in bringing the study of instinct back on to a firm scientific footing.[43]

Alfred Brehm inspired a new generation of ornithologists. Whatever his shortcomings, his voluminous and sentimental writings sparked an interest among his followers in what animals actually did. First among these was Oskar Heinroth. Director of the Aquarium in the Berlin Zoological Garden, Heinroth made the novel suggestion that instinctive actions like the courtship displays of ducks provide clues (just as anatomical structures do) to the evolutionary relationship between species.[44] The significance of this in the history of ornithology was monumental. For almost two centuries after Ray's death, the studies of bird systematics and bird behaviour had followed distinct and separate paths. By providing a conceptual link between these two areas of ornithology Heinroth's brilliant insight offered the possibility of unification. As often happens, someone else – in this case Konrad Lorenz – got the credit for the idea that behaviour could be a taxonomic tool – although Lorenz did acknowledge Heinroth as his inspiration.

The studies of Heinroth and Lorenz raised the possibility that the two separate strands of ornithology might one day be united. The person who succeeded in making this a reality was Erwin Stresemann. It is no coincidence that Stresemann recognised Lorenz's talents very early on and persuaded him to give up his medical studies in favour of behaviour. It was the right decision, for the discipline of animal behaviour brought clarity to centuries of controversy and confusion. Most of the early behaviourists' ideas about instinct have long been superseded, but their detailed observations and ingenious experiments on birds and other animals triggered a spectacular blossoming of behaviour studies that continue to this day.[45]

Lorenz's reputation has stood the test of time less well than those of his fellow Nobel laureates for several reasons, including the fact that he was a Nazi sympathiser during the Second World War and never properly apologised for this. Lorenz also clung tenaciously and arrogantly to his scientific ideas long after they were out of date; and, finally, the way he kept and studied animals in captivity is no longer considered scientific. Research has since become institutionalised and more professional. Yet, there is little doubt that through his close relationships with geese, jackdaws and other species Lorenz developed an unparalleled appreciation of their behavioural subtleties, an approach that in my view still has much to offer for those seeking a better understanding of bird behaviour.[46]

Niko Tinbergen's reputation has endured, in part because he helped to define the field by identifying the different types of questions one can ask about behaviour, referred to as Tinbergen's 'four whys', two of which are about mechanisms: (i) what causes the behaviour? (ii) how does the behaviour develop over an animal's lifetime? and two about evolution: (iii) its evolutionary history, and (iv) the adaptive significance of the behaviour. [47]

On hearing a great reed warbler singing one can ask first, what causes this behaviour? What are the internal mechanisms (hormones, neurones, musculature or brain function) and the external factors that cause the bird to sing? Second, one can ask why the great reed warbler sings in a particular way. What was it about that individual's development that made it sing in the way it does: was the song learnt from another great reed warbler or was its song 'innate'? This particular question focuses on the development of behaviour as a great reed warbler matures. Third, one can ask why the bird is singing in a particular way; why does a great reed warbler sing more like a sedge warbler than a wren? This is Heinroth's point about behaviour reflecting a bird's evolutionary history: the sedge warbler and great reed warbler sing in a similar way because they have a common ancestor. Finally, we can ask what the function of the behaviour is – how does singing increase the bird's likelihood of leaving descendants? This last question is about the adaptive significance of the behaviour – precisely what John Ray recognised as important in the seventeenth century, and what today has become the focus of an entire generation of researchers (including many ornithologists) and is now referred to as behavioural ecology.[48]

Tinbergen professionalised the study of animal behaviour. He made it scientifically respectable, replacing anecdotes with robust conclusions derived from ingenious field experiments or laboratory studies. It is a measure of how far the field of animal behaviour has developed

Singing is a remarkable and very obvious behaviour among birds, causing early ornithologists to ask how males, like this great reed warbler, acquire their loud and varied songs. What purpose does singing serve and what causes it? (From Howard's British Warblers, *1910)*

since Tinbergen's time that researchers now feel confident enough to go back and revisit some topics that were once considered either too tricky or taboo – notably, issues like intelligence and personality – exemplified by the construction and use of tools by crows.

The crows of New Caledonia routinely fashion hook-like tools from twigs or palm leaves to extract insect larvae from tree hollows. This is behaviour so sophisticated that it makes the great apes look almost dull-witted. In captivity, crows of New Caledonia presented with a straight piece of wire – something they are unlikely to have encountered before – can bend it into a hook to retrieve a piece of food. To test the birds' ingenuity, they are given a straight piece of wire about 10 cm long and an 'inaccessible' piece of meat in a small bucket inside a vertical clear Perspex tube. The crows quickly assess the situation, fashion the wire into a hook and use it to lift the bucket and retrieve the food.[49]

One explanation for the extraordinary cleverness of these crows is that they possess a brain module for making and using tools, little different from the modules that allow other birds to do other 'clever' things like string-pulling, constructing an intricate nest or knowing to migrate in a particular direction. Intriguingly, not all New Caledonian crows are able to solve the puzzles set them by researchers. In much the same way, not all goldfinches are able to perform the string-pulling trick: in captivity about a quarter of all goldfinches are 'inventors' and solve the string-pulling problem without ever having seen another bird do it; another quarter are 'imitators', who solve the problem only after seeing other birds perform it; the rest are 'duffers', who never learn the trick.[50]

These observations and simple experiments with New Caledonian crows reveal complex cognitive abilities and raise fascinating questions about how these birds acquire their tool-using skills. Studies of another tool-using bird, the woodpecker finch – one of Darwin's famous finches from the Galapagos Islands – provide some fascinating clues. Woodpecker finches use twigs or cactus spines (not hooks) to winkle insect larvae out of holes or from underneath bark. Not all adult woodpecker finches use tools – they exhibit the same kind of individual variation seen in New Caledonian crows and goldfinches. Intriguingly, however, when non-tool-using adult woodpecker finches are given the opportunity to watch tool-users, they fail to acquire the behaviour. In contrast, young woodpecker finches in captivity allowed to play with twigs and cactus spines *all* become tool-users regardless of whether they have seen other finches using tools. These remarkable results indicate that woodpecker finches acquire their tool-using skills through a genetic disposition for trial-and-error learning during a sensitive phase in their development.[51]

The differences in the abilities of individual crows, goldfinches and woodpecker finches could equally well be called differences in personality. Lorenz had noticed similar differences in his imprinted geese; some were shy, others were confident, yet at the same time he was blind to these differences, at least until they were pointed out to him, by the great evolutionary biologist Ernst Mayr:

> In 1951, Gretel [Mayr's wife] and I visited the Lorenzs at Buldern, Westphalia. There we had very long discussions, even controversies. At this period, Lorenz always talked about *the* greylag goose in a strictly typological sense. By contrast, I insisted that every greylag goose was different from any other one. 'If a greylag goose becomes widowed,' he said, 'he or she will never marry again.' I asked him on how many cases his statement was based, but he had only a vague answer. At any rate my insistence that every goose should be treated as an individual eventually resulted in Lorenz hiring a special assistant to keep track of the activities of every single individual in the flock. Each goose had its own card in the cardfile and all about each goose was recorded daily. Needless to say, within the first year he had already one or two cases of widowed geese remarrying. Many other sweeping statements about *the* greylag goose were likewise refuted by individual records.[52]

Even after this Lorenz never pursued 'personality' differences, perhaps because he was concerned that he might be accused of anthropomorphism, but today the study of animal behaviour has matured and

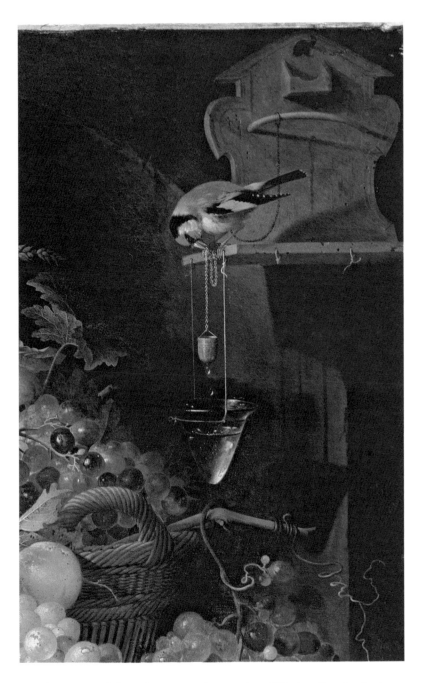

A favourite trick for at least two millennia: the goldfinch pulls a tiny bucket by a fine chain from a glass container in order to drink. (Detail from a painting by the Dutch-German seventeenth-century painter Abraham Mignon)

broadened its scope, and personality is suddenly a hot topic. Anyone that has lived or worked closely with animals knows that individuals often differ substantially in their attitudes, particularly towards people. One of the most striking, if somewhat unpleasant, demonstrations of this came from catching and keeping songbirds in the nineteenth century. The key to winning a bird show with a good-looking goldfinch or bullfinch was (and still is) to have a 'steady' bird – one that stands proudly and resolutely on its perch, staring confidently at all who pass in front of it. Steadiness can be improved by training, but all bird-keepers know that some individuals are inherently steadier than others. In the 1900s, when British bird-keepers took all their birds from the wild, they employed a simple but harsh procedure in order to identify the steadiest individuals. After a successful trapping session they would (allegedly) swing the cage around their head and then see which individuals were still standing on the perch. Those that were they kept; the rest they released. Bird-keepers probably never used the term 'personality' to describe this particular attribute, but they were well aware that individuals differed consistently in their behaviour. Monsieur Hervieux, who produced the first canary monograph in the early 1700s, wrote about the 'several dispositions and inclinations of canary birds', describing some as melancholy and some mischievous. Falconers also knew, as did pigeon and poultry fanciers, that the differences between individual birds could be considerable.

To most animal behaviour researchers from the 1930s onwards, such individual variation was inconvenient noise that obscured species-specific behaviours that were assumed to comprise average adaptive behaviours. Yet by the time Lorenz was undertaking his famous goose studies, psychologists had made entire careers out of studying personality in humans. Why, then, was it acceptable to study individual differences between humans but to ignore this in other animals?

I think the answer lies in the separate territories occupied by psychologists on the one hand and those studying animal behaviour on the other. For the latter, investigating something as nebulous as personality seemed subjective and a reversion to the anthropomorphism

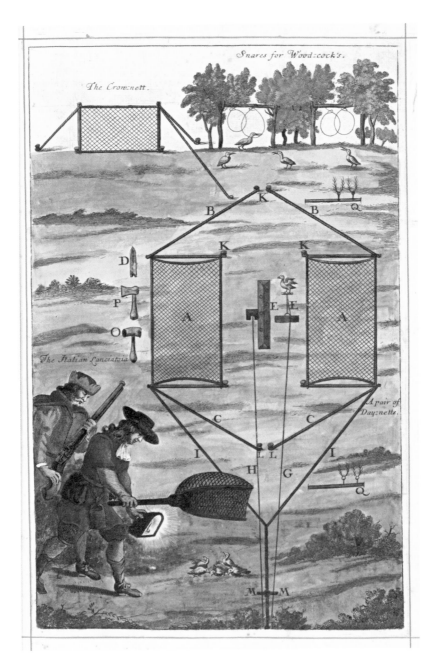

From their close encounters with birds, trappers knew that individuals of the same species could differ markedly in behaviour. This illustration, demonstrating different methods of catching birds, is from Samuel Pepys' coloured copy of The Ornithology of Francis Willughby *(Ray, 1678), redrawn from Olina (1622).*

they had struggled so hard to distance themselves from in the early days.

Psychologists often study personality by asking their subjects to assess themselves in terms of whether they are extrovert, talkative, energetic, sympathetic, shy and so on. It soon became clear that different traits tend to co-occur and people who considered themselves energetic also tended to be talkative, and vice versa. Psychologists refer to these combinations as 'personality dimensions', and because they seem to be universal across cultures they are thought to have a genetic basis. This has been difficult to test objectively in humans, but recent studies of non-humans, and birds in particular, show convincingly that personality traits are indeed heritable. This does not mean that environmental factors are not important; they undoubtedly are. Saying that personality is 'heritable' means only that it has a genetic component.

A team of researchers in the Netherlands Institute of Ecology has assessed whether great tits are bold or shy – a simple personality trait easily measured by seeing how quickly individuals approach a novel object in a novel environment. Assessment involves placing a bird in an aviary with something it has never seen before, such as small plastic toy, and recording how long it is before the bird comes to look at it. Not only are the differences in this exploratory behaviour between individual great tits clear-cut and consistent over time, but artificial selection breeding experiments in captivity show that the differences are heritable.

In turn, this suggests that these personality types have some adaptive value otherwise all individuals would have converged on a single 'optimal' personality over evolutionary time. The beauty of the great tit system is that wild birds can be brought into captivity for a personality test that takes just a few hours, before being released back into the wild, where their breeding success and survival are monitored. This ingenious study showed that year-to-year variation in ecological conditions was responsible for maintaining the different personality types. When winter food such as beech mast was abundant bold males and shy females survived best. But when food was scarce shy males and bold females had better survival. The sex difference is linked to the fact that male great tits are always dominant over females. In years of abundant winter food, competition for food is relaxed and all birds survive well, resulting in strong competition among males for territory the following spring. Under such circumstances bold males do better simply because they are more aggressive and more likely to secure a breeding territory. Females, on the other hand, can relax; with little competition for abundant winter food they avoid conflict and all its accompanying stresses and strains. When winter food is scarce, on the other hand, mortality is high and shy males do best because competition for territories is relaxed. For females, however, being feisty helps to ensure survival when food is scarce.[53]

Species as well as individuals also differ in their temperament or personality and two groups of birds stand out as being particularly 'clever': crows and parrots. What is special about crows and parrots? The answer is that they live in complex environments where behavioural flexibility is the key to survival. In fact, in terms of their behaviour crows and parrots are more like primates – monkeys and apes – than birds. They face similar ecological problems, occupying demanding environments and feeding on a wide range of food types. They also live in complex social groups and have long periods of dependence on their parents. And just like primates, corvids (crows, magpies and jays) and parrots also have relatively large forebrains and more sophisticated cognitive skills and behavioural flexibility than any other group of birds.

Perhaps the most famous parrot of all time was Alex, an African grey who died while I was finishing this book in September 2007, at the age of thirty-one. Alex was owned and trained by Irene Pepperberg, who started her academic career as a chemist at Harvard but switched to the study of animal communication and is now justly famous for her remarkable studies of the cognitive abilities of parrots. The beauty of working with parrots is that they can be trained to talk and reveal what they are thinking. Pepperberg worked with Alex for over twenty years, and while his vocabulary was not exceptional, he was exceptional in that, unlike most parrots, Alex usually said what he meant.

The African grey parrot is much more than a mimic of the human voice. The most famous African grey of all time, Alex, had the cognitive abilities of a four- or five-year-old child. (From Frisch, 1743–63)

Pepperberg trained Alex to identify over one hundred types of object by their colour, shape and material. Presented with a tray of objects of different colour, shape and number and asked, for example, how many green balls there were, he answered correctly, demonstrating that he understood the concepts of colour and shape and was able to link the two together. By altering the number, type and colour in each test, Pepperberg showed clearly that Alex understood this kind of problem. No other species, bird or mammal, can do anything as sophisticated. The intellectual ability of Alex was similar to that of a human four- or five-year-old, but emotionally, she says, Alex had the 'negative, self-centred behaviour of a two- to three-year-old', which is why there are 'so many abandoned parrots'.[54] The parallel with human infants is compelling: children typically go through several cognitive stages relating to their ability to keep track of an object or an individual that moves out of sight, and hence being able to retain a mental image of the hidden object. Parrots excel at this, as do crows when they hide and recover food items.

Birds that cache food are remarkably good at remembering where they have to look to recover it; the Clark's nutcracker (another member of the crow family), for example, appears to be able to remember hundreds of locations where it has concealed food. Even more remarkable is the ability of some corvids to understand that certain food items are more perishable than others and only worth eating if recovered within a certain time. Nicky Clayton at the University of Cambridge tested the abilities of American scrub-jays and found them to have an uncanny ability to recover different items after appropriate time periods, indicating that they have mastered the concepts of 'where', 'what' and 'when'. But they can do more, much more. Scrub-jays sometimes pilfer the caches of other jays by remembering where they have hidden food. Birds that knew they had been watched hiding food by another jay subsequently re-hid it, but only if they had been a pilferer themselves. Birds with no personal pilfering experience never re-hid food when they had been watched. These remarkable results indicate that scrub-jays can also remember the social context

Members of the crow family, known as corvids, are considered among the most
'intelligent' of birds. These four North American jays, painted by Alexander Wilson
(from Wilson and Bonaparte, 1832), are the blue (top left), Steller's (middle),
Florida scrub (lower) and Canada jay (top right).

of their previous behaviour and then adjust their current behaviour to minimise the risk of having their food stolen – suggesting that they can plan ahead, transferring their knowledge acquired as a pilferer to being pilfered! Furthermore, they also remember which particular bird was watching when they hid a particular cache and take protective action accordingly; protecting only those food caches that particular competitor saw them hide. In contrast, when their life-long partner watched them cache they did not bother to re-hide the food, because partners usually share their food caches and jointly protect them from potential pilferers.[55]

We are forced to conclude that birds like parrots and crows are pre-adapted to learn certain things – especially those relevant to their survival. These recent studies of bird behaviour finally allow us to make sense of all those curious incidents that early ornithologists felt so passionately must indicate conscious thought. They also demonstrate very effectively that there is no nature–nurture divide: no hard and fast distinction between behaviours that are innate and those acquired through learning. Instead there is a continuum of behaviour dictated and moulded by the environment. In simple undemanding environments innate responses or basic trial-and-error learning are sufficient to allow birds to survive and thrive. But at the other extreme, for those species with complex social lives and occupying environments where finding food is tricky or requires the use of tools – like acorn wood-peckers, corvids and parrots – natural selection has favoured bigger brains and greater behavioural flexibility.

The near-perfect match between a bird's behaviour and the world in which it lives that so impressed John Ray and others is the outcome of a complex series of interactions between its genes and its environment, a product of natural selection, rather than God's providence. And as we will see in the following chapter, there is no better evidence for this than migration.

Most of the white storks that breed in Europe migrate to Africa for the winter.
This image is from John Gould's The Birds of Great Britain *(1873).*

4

Disappearing Fantasies

The Emergence of Migration

I'm sitting in a medieval castle, a small one, on the outskirts of the sleepy village of Radolfzell in southern Germany. It is cold outside and a low winter sun is casting long shadows from the apple trees in the surrounding orchards across the frosty ground. Unlikely as it might seem, the castle houses a famous institute of bird migration.[1] In front of me is a computer screen, and on it a pixellated outline of the Dark Continent. From the very centre of Africa a point no bigger than a full stop is flashing at the same frequency as my heart beats. It is almost as though it *is* a heartbeat, but the signal is from a radio carried by a white stork standing motionless on the hot savannah. Five months previously, before this particular bird left its nest in eastern Germany, it was fitted with a small transmitter that is now monitored several times a day by a satellite streaming across the heavens, pinpointing its exact location, its body temperature and whether it is flying, walking or standing. The stork is out of sight of almost any human and I feel overawed and humbled by so effortlessly being able to 'see' this bird. A couple of mouse clicks and I can re-create its southward journey; its day-to-day progress across eastern Europe, the Bosporus, Turkey, Eilat, Sudan and into Chad where it has been for several weeks during the European winter. Another mouse click, and I can look at another stork. Ringed in the same part of Germany, this one continued to fly south and is wintering near Cape Town and by the time it returns to Europe next spring it will have flown some 24,000 km (just under 15,000 miles). The technology that allows me this view of a bird's life is a miracle, but a tiny one compared with migration itself. So far the satellite transmitters are suitable only for large birds – albatrosses, eagles, swans and storks – but soon, as the technology improves, tiny transmitters powered by the sun will allow us to track swallows, swifts and martins on their epic journeys across the Sahara Desert and back.

When John Ray wrote *The Wisdom of God* he was in no doubt about the fact of migration; what fascinated him was *how* and *why* it occurs:

But how come they to be directed to the same place yearly, though sometimes but a little island, as the soland goose [gannet] to the Basse of Edinburgh Frith [sic], which they could not possibly see, and so it could have no influence upon them that way? The cold or the heat might possibly drive them in a right line from either, but that they should impel land-birds to venture over a wide ocean, of which they can see no end, is strange and unaccountable: one would think that the sight of so much water and present fear of drowning should overcome the sense of hunger, or disagreeableness of the temper of the air. Besides, how come they to steer their course aright to their several quarters, which before the compass was invented was hard for a man himself to do, they being not able, as I noted before, to see them at that distance? Think we that the quails for instance, could see quite cross the Mediterranean Sea? And yet, it's clear, they fly out of Italy into Africa, lighting many times on ships in the midst of the sea, to rest themselves when tired and spent with flying. That they should thus shift places, is very convenient for them, and accordingly we see they do it; which seems to be impossible they should, unless themselves were endowed with reason, or directed and acted by a superior intelligent cause.[2]

Today, Ray's assumption that gannets, quail and warblers migrate seems no more than common sense, but migration was far from universally accepted in the 1600s. When birds disappeared at the end of the summer it was commonly believed that, rather than setting off across the sea, many simply went into hibernation, hiding themselves in crevices or in the mud at the bottom of ponds. This is rather curious given that migration had been known, or at least assumed, for several hundred years before Christ.

For certain species migration was fairly obvious. White storks disappeared in winter and reappeared the following spring, and because they are large and often travel in huge, conspicuous flocks, the ancients could actually see them migrating. The birds' size and confident manner probably dispelled any doubts about their ability to cross the sea successfully.

And it wasn't just large birds. A poem from the sixth century BC attributed to the Greek poet Anacreon and translated by Thomas Stanley during Ray's lifetime confirms an early belief in migration:

> Gentle swallow, thou we know
> Every year dost come and go,
> In the Spring thy nest thou mak'st
> In winter it forsak'st,
> And diverst'st thyself awhile
> Near the Memphian Towers or Nile.[3]

The 'Memphian Towers' refer to the ancient Egyptian city of Memphis; the Nile is another reference to the fact that swallows went from Greece to Africa.

Three centuries later, Aristotle reaffirmed the common belief in migration:

For all animals have an instinctive perception of the changes of temperature, and, just as men seek shelter in houses in winter, or as men of great possessions spend their summer in cool places and their winter in sunny ones, so also all animals that can do so shift their habitat at

Overleaf: *White storks, cranes and ducks migrating over the sea, from Frederick II's* Art of Falconry, *produced in the thirteenth century. (From Sauer et al., 1969)*

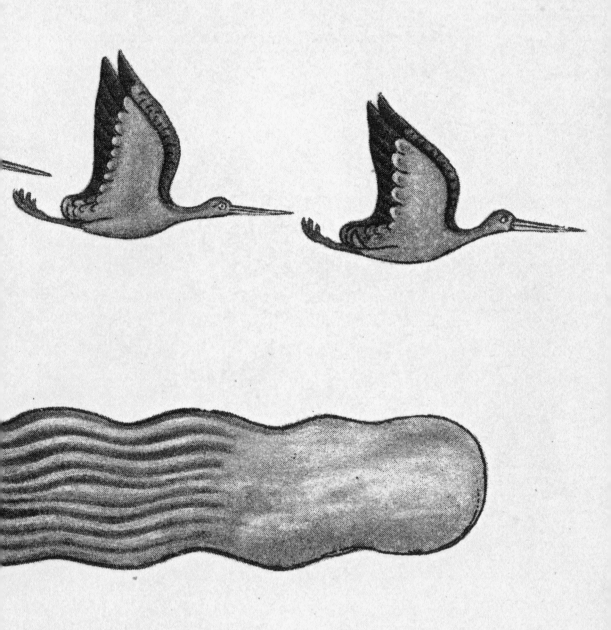

various seasons … others migrate … quitting … the cold countries after the autumn equinox to avoid the approaching winter, and after the
spring equinox migrating from warm lands to cool lands to avoid the
coming heat.[4]

For millennia the seasonal appearance and disappearance of birds has
been obvious to hunters and casual observers alike. The swallow, the
stork and the nightingale were harbingers of spring, but where most of
them went to or came from was a subject of great speculation. Sometimes, as when the Israelites were miraculously saved from starvation
by a timely fall of migrating quail, their appearance became imbued
with great religious significance: 'It came to pass that at even the quails
came up and covered the camp.'[5]

Aristotle's view of migration seems to have been remarkably modern, so where did the idea of hibernation come from, and why did it
take two thousand years to dispel?

The answer to the first question is that Aristotle himself was to
blame. In his *History of Animals*, having made a clear statement about
migration, he adds:

> A great number of birds also go into hiding; they do not all migrate, as
> is generally supposed, to warmer countries … Swallows, for instance,
> have been often found in holes, quite denuded of their feathers … And
> with regard to this phenomenon of periodic torpor there is no distinc-
> tion observed, whether the talons of a bird be crooked or straight; for
> instance, the stork, the ouzel, the turtle-dove, and the lark, all go into
> hiding. The case of the turtledove is the most notorious of all, for we
> would defy any one to assert that he had anywhere seen a turtle-dove
> in winter-time; at the beginning of the hiding time it is exceeding-
> ly plump, and during this period it moults, but retains its plumpness.
> Some cushats [pigeon] hide; others, instead of hiding, migrate at the
> same time as the swallow. The thrush and the starling hide; and of birds
> with crooked talons the kite and the owl hide for a few days.[6]

It is not hard to see why Aristotle and other naturalists were confused. They knew that certain mammals, including the bat – considered a type of bird until the sixteenth century – overwintered in a state of torpor and could be roused to their normal state by the application of warmth.[7] Finding the occasional torpid or dead bird hidden away in a crevice must simply have reinforced the idea that hibernation was how certain birds, just like bats, spent the winter. Also, as with all myths this one contained a grain of truth. Some birds, including swallows and swifts we now know, can indeed go torpid, albeit for short periods, when food is scarce or the weather cold, and can be restored by warmth.[8]

In the thirteenth century the Holy Roman Emperor, Frederick II of Hohenstaufen – a keen falconer – had Aristotle's writings translated into Latin, but chose to ignore his comments on hibernation, just as he ignored the Bestiaries, the standard natural history texts of the day. More provocatively, Frederick also ignored the philosophy of the Church and instead based his knowledge of natural history on direct observation and inductive reasoning, rather than what was ordained. Having seen migration for himself, he was in no doubt about its occurrence.

Frederick's view on the seasonal appearance and disappearance of one particular migratory species, the barnacle goose, reveals his common sense. This species was thought to appear spontaneously from timber floating in the sea – a rumour started by the Welsh medieval cleric Giraldus Cambrensis, nephew of David, Bishop of St David's, who acquired this story during a visit to Ireland with Prince John in 1185–6. In his *Topographica Hibernica*, Giraldus wrote:

There are in this place many birds which are called Bernacae: Nature produces them against Nature in the most extraordinary way. They are like marsh geese but somewhat smaller. They are produced from fir timber tossed along the sea and are at first like gum. Afterwards they hang down by their beaks as if they were seaweed attached to the timber, and are surrounded by shells in order to grow more freely. Having thus

Frederick II of Hohenstaufen, Holy Roman Emperor, was a keen falconer.
His treatise on the subject contained many excellent observations and
inferences about the lives of birds. (From Sauer et al., 1969)

in the process of time been clothed with a strong coat of feathers, they either fall into the water or fly freely away into the air. They derive their food and growth from the sap of the wood or from the sea, by a secret and most wonderful process of alimentation. I have frequently seen, with my own eyes, more than a thousand of these small birds, hanging down on the sea-shore from one piece of timber, enclosed in their shells and already formed. They do not breed and lay eggs like other birds, nor do they ever hatch any eggs, nor do they seem to build nests in any corner of the earth. Hence Bishops and religious men in some parts of Ireland do not scruple to dine off these birds at the time of fasting, because they are not flesh nor born of flesh.[9]

Frederick considered this complete claptrap. Instead, he said, the geese simply breed in distant lands and then appear at the end of the breeding season.[10] The barnacle story was too convenient to be ignored, however, since it allowed the clergy to eat goose during Lent. Over time the story was embellished and reinforced by fantastic illustrations, so much so that in the 1500s William Turner wrote to tell Conrad Gessner that he was (reluctantly) convinced by it. Gessner in turn included it in his encyclopaedia, as did Aldrovandi, and John Gerard in his *Herball* of 1597 told how he had found birds covered in soft down inside the barnacle shells, and included an illustration showing tree-geese hatching from barnacles. In repeating the story these natural history writers were simply bowing to tradition. With no evidence to the contrary, denying the existence of barnacle geese would have seemed like an omission.

Interestingly, Albert the Great doubted the story: 'They also say that these animals are sometimes generated from rotten logs in the sea ... maintaining that no one has ever seen these birds copulate or lay eggs. Now this is entirely absurd, for, just as I have said in preceding books, I and many of my friends have seen them both copulate and lay eggs as well as nourish their young.'[11] This is odd, for while Albert is right about the barnacle story being absurd, he was obviously mistaken about seeing barnacle geese breeding. It was precisely

Overleaf: *Two pages from the original manuscript of Frederick II's* Art of Falconry, *with its beautiful miniatures, held in the Vatican Library. (From Sauer et al., 1969)*

bistarda

anas campestris

grus

coturnix

perdix

ayro

upupa

pica

rapaces sub se conprimant. ut
bistarde. et que dicuntur anates
campestres similes sunt bis-
tardis. s. longe minores. et in
estate turpem sonum faciunt
ad desiderium coitus. Per
uolatum suum multiplicr. se
defendunt aues. Nam alie p
longum uolatum querunt e-
uadere ut grues. alie p uelox
ad ueniendum ad locum defensionis sue. ut pdicam. et cotur-
nicum modi. alie p diuturni-
ta et per cessiones quas faciunt in uolando. ut modi ay-
ronum. cornices. upupe. uan-
nelli. pice. et plures alie. Alie ui-
uo p uolatum ad altiora se querunt defendere. et hoc duobz
modis. aut enim directe ascendunt. ut columbi. turtu-
res anates campestres. aut
in gyrum eundo uolando ascendunt ad defensionem sui. ut
ayrones qui scim q dictum est du-
plici utuntur defensione. p uolatum diuturnum scilicet et
ascensum. et omnis q per uolatu
querunt defendere se ascenden-
do. ideo ascendunt qd non
possunt superari. et uincu ab aui-
bus rapacibz in magis ascende-
do. Alie sunt que querunt defendere se uolando uersus loca
de quibz timent rapaces quam

uis ad ea non descendunt. ut
anseres anates. et pluuer alie
de minera que uolant circiter
loca in quibz sunt aq magne
nemora calami. et canne de
quibz locis timent aues rapi-
ces. ad hmodi enim loca timent
descendere et accedere. Alie ad
maiorem securitatem sui uolatum sui faciunt i crepuscu-
lis. et in nocte. ut noctue. bu-
bones. et luuerzin. qui p eo q
sunt timorosi nocte uolant
securius. De modo defen-
sionis. E modo autem sionis
defensionis quem hnt
aues refugiendo ad lo-
ca securiora dicendum e. qr u-
niuersalit omnis aues ad plus
si possunt refugiunt ad loca
natiuitatis sue. ut ad similia
illis. Ille siquidem q nate sunt
apd aquas ad eas confugiunt
quam qdam natando in eas
solum hnt defensionem ut
pellicani. qdam submergendo
se penitus sub aquis. ut modi mergorum. anarum. et a-
liarum plurium. qdam non
penitus submergendo se. sed
in parte ut ciconie. anseris
modi. Aues uero que no na-
tant nec sunt aquatice. timo-
re auium rapacium ad aqua
confugiunt. facunt enim q

uanellus

columbus

anser

pelicanus

magnus

anates

ciguus

aues rapaces & aquis timet
& circa aquas debilissime sut
ut dicm e in capitulo de diuisi
one auium· plures itaqz a
uium ad aqs confugiunt quo
dam p defensione sua· quedam
p cibo· quedam p utroqz. Ille
uero q nate sunt int arbores
ad arbores confugiunt· ut
modi corniciu· pici· galli
or plium. Et ille q nate sic
ppe aqs· & iste eedem quq in
ter arbores confugiunt· quo
ad aqs qn ad arbores ut mo
di ayronu. Si uero nate q
sup prata· fructices· aut du
mos ad ea confugiunt· ut
modi· sturnelli & auiculariu
plures· q nate sunt in rupibz
ad saxosa confugiunt ut ra
paces· sed q nate sunt sup
tram & sunt coloris terestri
dum latitant & sic se re
comendant· ut perdices· co
turnices· cosardi· alaude· ca
pestres· calandre & auicule
plures· de quibz multe sue
adeo stolide in cautela sui q
ardentes se esse sentiunt ter
ra capiunt· etiam manibus
hominu & qn insequit eas
rapar ad tram confugiunt
    Perdices fasiani· & p mo
du qn sunt curti uolatus & p
adiutoriu loci q runt defen

sionem sibi nuc libent recedut
longe a loco apto defensioni
sue· defensionibz & pugna
tionibz pdictis generalit uuut
maior pars auium· q aut plu
ribz· quidam paucioribus·
Spetiales autem defensiones
insunt auibz· & ppie ut bis
tardis & anatibz· campestriu·
est ppria defensio emittere lon
ge a se stercus sui in aues
rapaces que psequuntur eas.
    Rursus bistarde & anates
campestres cont aues rapa
ces horripilant plumas agri
fando se· & eleuant alas depo
nendo caput ad modu gallor
pugnantiu quod pp timo
rem faciunt· & tamen bistar
de & anates campestres cont
aues rapaces· persequunt alis
& pectore aues rapaces· Sut
& alie aues que refugiut ad
societatem & congregatione
aliar sue spetiei ut int eas
securiores sint· & p eas defen
dant sicut sunt columbi· gru
es· sturnelli· & alie fere om
nino etiam tota agmina in
se ipa densius se constringut
auibz rapacibz facientibz in
sultum. Et ea q aures in socie
tate tutiores sint signum e
q ples sunt spes auium q
concurrunt ad defendendu

bistarda

bistarda

anas campe
stris

grus

pica

cornix

columbi

cornix

ayro

turdi

sturnel
lus

perdix
cornix

fasianus

perdix

because no one had ever seen their northerly breeding grounds that the myth persisted and it was only when the Dutch explorer Gerrit de Veer found the birds breeding on Spitzbergen in 1596 that the story was at last dispelled.[12] Frederick's idea of a remote breeding location was finally confirmed, but because his writings lay undiscovered until the 1780s his sensible views played no part in the ongoing discussions about where birds went in winter.[13]

With the exception of the barnacle goose, the various authors of the medieval Bestiaries had no problem with migration. Of the swallow, it: 'flies across the sea. And there overseas it lives during the winter'; of storks: 'messengers of spring, these brotherly comrades, these enemies of serpents, can migrate across oceans and, having collected themselves into column of route, can go straight through to Asia'; of quail: 'When the summer is over they cross the sea.' Albert the Great reinforces the idea of migration by criticising those who said storks winter somewhere in the east and that 'there is a plain in Asia in which they wait for each other', saying that this cannot be true for

Belief that barnacle geese grew on trees or sprang from logs floating in the sea persisted for centuries in part because it allowed the clergy to eat goose during Lent.
(From a thirteenth-century Bestiary)

there is no value of going east because the climate is the same. Of the quail he says: 'Many believe about this bird that when it goes away it goes across the sea.'[14]

So far so good. But then, in a careless gesture, Albert fans the dying ember of hibernation, referring to the hoopoe as 'a familiar bird that sleeps in winter like the bat'.[15]

The ember smouldered unnoticed for almost two hundred years until Olaus Magnus, Archbishop of Uppsala, in Sweden, ignited in the 1500s what was to become one of the longest-running debates in the history of ornithology:

> Although the writers of many natural things have recorded that the swallows change their stations, going, when winter cometh, into hotter countries; yet, in the northern waters, fishermen oftentimes by chance draw up in their nets an abundance of swallows, hanging together like a conglomerated mass.[16]

Images of swallows in the mud appealed to some of the great early naturalists,[17] but because others were sceptical, in the 1660s the Royal Society of London commissioned one of its members, the Polish astronomer Johannes Hevelius, to establish 'What truth there is ... concerning swallows being found in winter under waters congealed, and reviving, if they be fished [out] and held to the fire.' Hevelius's surprising answer was that:

> It is most certain, that swallows sink themselves towards autumn into lakes ... many has assured him of it, who have seen them drawn out with a net together with fishes, and put to the fire, and thereby revived.[18]

The source of this erroneous information was one of Hevelius's correspondents, Johannes Schefferus, professor at the University of Uppsala, still loyal to his countryman Olaus Magnus and to his alma mater. Recounting this story a century later, Buffon was incensed by it, not least because, knowing that Hevelius favoured the submersion hypothesis over migration, he felt that the Royal Society should have chosen someone more independent (and a naturalist rather than an astronomer) if they wanted a reliable answer.[19]

Endorsed by the Royal Society, the view that swallows, along with swifts and martins, spent their winters under water became increasingly entrenched. Sucked into the debate, others claimed to have witnessed the phenomenon and seen swallows taken from their watery resting places.[20]

Genuine ornithologists, however, were unconvinced. Belon knew from first-hand experience that birds migrated, as did Sir Thomas Browne, who, in his notes on the birds of Norfolk, written in the 1660s, said: 'Besides the ordinary birds which keep constantly in the country, many are discoverable both in winter and which are of a migrant nature and exchange their seats according to the season. Those that come in the spring coming for the most part from the southward, those which come in the autumn or winter from the northward.'[21] A

A draught of fishes and swallows. The idea that swallows overwintered under water was fixed in people's minds by this woodcut which accompanied Olaus Magnus's erroneous account of 1555. Here we see fisherman on the ice of a Scandinavian lake with a net full of both fish and fowl. (From Olaus, 1555)

physician by profession, Browne was well known for his no-nonsense approach to natural phenomena and a regular correspondent of John Ray. None the less, when writing the *Ornithology* in the 1670s Ray was still rather undecided about migration, at least where the swallow was concerned, and opted for an inelegant compromise that covered all possibilities:

> What becomes of swallows in winter time, whether they fly into other countries, or lie torpid in hollow trees, and the like places, neither are natural historians agreed, nor indeed can we certainly determine. To us it seems more probable that they fly away into hot countries ... and then that either they lurk in hollow trees, or holes of rocks and ancient buildings, or lie in water under the ice in northern countries, as Olaus Magnus reports.[22]

Thirty years later, however, in *The Wisdom of God*, Ray had abandoned the idea of torpor:

> The migration of birds from an [sic] hotter to a colder country, or a colder to an hotter, according to the seasons of the year, as their nature is, I know not how to give an account of, it is so strange and admirable. What moves them to shift their quarters?[23]

The view that swallows and some other birds remained torpid or under water persisted well into the eighteenth century and even the great Linnaeus believed it, at least until he was taken to task over it by one of his students.[24]

Before scoffing at such naïvety, we should remember that scientific knowledge at this time was still rudimentary; bloodletting was routine and many people still believed in unicorns and dragons, so a belief in hibernating birds was hardly exceptional.[25]

In fact, by the mid-1700s it was generally accepted that the majority of birds that disappeared in autumn did migrate. The swallow and similar species (martins and swifts), however, remained a special case, precisely because, on the face of it at least, there seemed to be so much evidence in favour of hibernation. None the less, the debate was coming to a head and was played out by three men.

The first is the Right Honourable Daines Barrington, English lawyer and Fellow of the Royal Society, absolutely convinced that swallows do not migrate. For him it simply was not logical that swallows should attempt anything so dangerous.[26]

The second is Thomas Pennant, a well-travelled, wealthy magistrate and author of the much-respected *British Zoology*. Pennant was both a bird enthusiast and a fan of John Ray, having owned a copy of the *Ornithology* from the age of twelve. He had more practical experience of birds than Barrington and, perhaps as a result, was a firm believer in migration.[27]

The final member of this curious trio is Gilbert White, a gentle country curate – probably not dissimilar to John Ray in manner – whose *Natural History and Antiquities of Selborne*, based on White's correspondence with Pennant and Barrington and first published in 1789, eventually became one of the best-selling books of all time.[28]

They are an odd trio. Barrington the lawyer could argue black was white, or, worse, that torpor was more plausible than migration. Pennant, unmovable in his belief in migration; and White dithering uncertainly in the middle. It wasn't a three-way discussion, however; the correspondence was predominantly between White and the other two.

Barrington's philosophy was that biological puzzles like migration could be solved through the power of logic:

> Though it might be answered, that it is not necessary, those who endeavour to shew the impossibility of another system or hypothesis, should ... be obliged to set up one of their own; yet I shall, without any difficulty, say that I at least am convinced swallows (and perhaps some other birds) are torpid during the winter.[29]

It became Barrington's personal mission to take the evidence for migration and demonstrate just how illogical it was. For Pennant, on

The hoopoe: 'a familiar bird that sleeps in winter like the bat', according to thirteenth-century cleric and polymath Albert the Great. This image is by Johann Walther from around 1650.

the other hand, the appearance of swallows on ships out at sea was evidence enough. Hadn't Christopher Columbus on his second voyage seen a swallow while still ten days from St Domingo in the West Indies? And what about the French botanist Michel Adanson, who, in his *Voyage to Senegal*, describes how some fifty leagues from the coast of Africa four swallows alighted on the shrouds of the ship, were captured and identified as English swallows. And Sir Charles Wager, First Lord of the Admiralty in the early 1700s, who reported how on entering the English Channel a great flock of swallows settled on the rigging, such that: 'Every rope was covered with them: they hung on one another like a swarm of bees: the decks and carvings were filled with them: they seemed spent and famished, and were only feathers and bone, yet after a night's rest they resumed their flight.'[30]

Compelling evidence for migration one might think, but Barrington rejects it and with wonderfully perverse logic: 'It seems, therefore, that birds are by no means calculated for flights across oceans, for which they have no previous practice: and they are, in fact, always so fatigued, that, when they meet a ship at sea, they forget all apprehensions, and deliver themselves up to the sailors.'[31] Barrington continues, accusing Adanson of carelessness and for failing to check whether the swallows he saw were European swallows. In fact, Barrington says, if Adanson had looked closely he would have seen that they were African swallows merely flitting from one headland to another. And, in a final flourish, Barrington adds that if swallows migrated there would be more sightings at sea rather than just a handful.

The migration lobby retaliated by arguing (correctly, of course) that the swallows that end up on ships were those that have encountered unfavourable conditions and are exhausted and that, normally, migrating swallows fly high and possibly at night – hence the infrequent sightings. Barrington has an answer for this, too, sarcastically pointing out that birds may indeed fly high. After all, he said, hadn't Charles Morton in his *Harleian Miscellany* suggested they go to the moon in winter?[32] Moreover, after carefully studying how high birds fly, Barrington concludes: 'I doubt very much whether any bird was ever seen to rise to a greater height than perhaps twice that of St Paul's cross and the rising to … extraordinary height … is destitute of proof.' He also asks, wouldn't the 'rarefaction' of the air at high altitude be inconvenient for respiration, adding further that if birds could fly high, they would do so at other times – but they don't. As for them migrating at night, that was ridiculous because, as everyone knew, birds sleep at night and if disturbed were usually disorientated in the dark: 'it is therefore inconceivable that they should choose owl-light for such a distant journey'.

It is now known – primarily from studies using radar during the Second World War –that many small birds do migrate at night and at relatively low altitudes, around 2,000 feet (700 m). What is more, the phenomenon of 'visible migration' which occurs during the day is also well established.[33]

Barrington's carefully crafted arguments *against* migration are often convincing. His arguments *for* torpidity are, by contrast, pathetic. He even admits: 'I have not, I must own, myself ever seen them in this state [of torpidity]; but, having heard instances of their being thus found, from others of undoubted veracity, I have scarcely the least doubt with regards to this point.' He adds: 'Why is it more extraordinary that swallows should be torpid during the winter, than that bats are found in this state?'

Barrington singularly failed to address Buffon's question of how swallows survive for six months under water,[34] or the fact that several researchers, including John Hunter, had plunged swallows under water only to have them expire within a few minutes.[35] Barrington also rejected the results of Buffon's experiment in which swallows were placed in an ice-house to see if they would go torpid. They didn't, but in typical form Barrington dismisses this, too: 'The very name ice-house almost strikes one with a chill: I placed however, a thermometer in one near Hyde Park Corner, on the 23rd November, where it continued 48 hours, and the mercury then stood at 43.5 by Fahrenheit's scale.' This simply wasn't cold enough: he had even seen swallows flying around at this sort of temperature. What's more, Barrington sneered, Buffon conducted his experiment at the wrong time of year, when the birds weren't ready to go torpid.

In a final piece of damning evidence designed to appeal to his readers' logic, Barrington says that it is well known that swallows breed in Lapland: if these birds migrate south to Africa what possible reason could there be for them to return to Lapland when there is all that suitable habitat in between? The answer, he claims, is that they do not migrate![36]

Barrington seems to have had an answer for everything. Richard Mabey sums him up as having a 'whiff of self righteousness about him – as if he was constantly disappointed by nature's failure to live according to the tidy moral scheme he thought appropriate for it – and that his care, his passion, was for the theory not the creatures'.[37] A *whiff* of self-righteousness? Barrington was downright arrogant and, like many clever people, ultra-critical of other people's ideas and hopelessly uncritical of his own.

Thomas Pennant was just as confident that migration *did* occur, but presents his case rather more concisely, writing in his *British Zoology*:

> Of the three opinions [migration, torpor or submersion], the first has the utmost appearance of probability; which is, that they remove nearer the sun, where they can find a continuance of their natural diet, and a temperature of air suiting their constitutions. That this is the case with some species of European swallows, has been proved beyond contradiction ...[38]

Stuck midway between his two correspondents, Gilbert White is still not sure whether swallows migrated or not. Writing to Pennant on 28 February 1769 he says:

> When I used to rise in a morning last autumn, and see the swallows and martins clustering on the chimneys and thatch of the neighbouring cottages, I could not help being touched by a secret delight, mixed with some degree of mortification: with delight to observe with how much ardour and punctuality those poor little birds obeyed the strong impulse towards migration, or hiding, imprinted on their

No birds caused more debate over their disappearance each winter than swallows and martins. Long after it was recognised that other birds migrated, swallows and martins were thought to hibernate. (Painting by George Edward Collins)

minds by their great Creator; and with some degree of mortification, when I reflected, that after all our pains and inquiries, we are yet not quite certain to what regions they do migrate; and are still farther embarrassed to find that some do not actually migrate at all.[39]

And in February 1771 in a letter to Barrington:

> You are, I know, no great friend of migration; and the well attested accounts from various parts of the kingdom seem to justify you in your suspicions, that at least many of the swallow kind do not leave us in the winter, but lay themselves up like insects or bats, in a torpid state, and slumber away the more uncomfortable months ... But we must not, I think, deny migration in general; because migration certainly does subsist in some places, as my brother in Andalusia has fully informed me.[40]

White's brother John had been sent down from Oxford in July 1750, for being party to a wedding between a gentleman commoner and the daughter of an innkeeper – of the Lamb at Wallingford, a pub of 'no good character' – and later entertaining the bride and her sisters in his college rooms. Expulsion from Corpus Christi meant that instead of securing a living through his college John was forced to look further afield, choosing to become chaplain to the garrison on Gibraltar, which is where he and his wife went in 1756. Over the following fifteen years John had unprecedented opportunities to witness migration directly and to tell Gilbert of his observations. Other than seeing swallows and other birds winging their way across the Strait of Gibraltar for himself, White could hardly have had more compelling evidence for migration.[41] Yet he persisted in the belief that swallows hibernated. The *Natural History of Selborne* is riddled with references to hibernation, with White recounting year after year how he has seen swallows on mild days early in the season: '... it is reasonable to suppose that two whole species, or at least many individuals of those two species, of British hirundines [swallows and house martins], do never leave this island at all, but partake of the same benumbed state ...' When they first appear, he says, they do so near water and if the weather turns cold 'they immediately withdraw for a time'. This, he suggests, is more consistent with hibernation: 'it makes sense that they would retire to their hybernaculum, close at hand rather than returning to warmer climes'. White doesn't consider the possibility that these early arrivals caught out by bad weather simply die!

Why should White, so astute in all other aspects of natural history, cling so tenaciously to hibernation? Richard Mabey has the correct answer, I think, suggesting that White's interest in swallows, martins and swifts was more emotional than scientific. White loved swallows – their sheer visibility at the nest and apparent fearlessness of man were endearing; he simply wanted it to be true that they hibernated, so that during the dark days of winter he could imagine them close by.[42]

By perpetuating doubts over migration, White's *Natural History of Selborne* prolonged the controversy, allowing it to rumble on into the nineteenth century. Assessing the evidence once again in the early 1800s Thomas Forster, an English naturalist and astronomer, asked 'whether the few species of swallow which visit us in spring, and retire in autumn, and ... are of a nature quite different [from other migratory species], and becoming torpid during winter, is the question ...'. He continues: 'It is more difficult to reconcile their opposite opinions and evidence, by supposition that some species migrate, and others lie torpid, than to suppose that accidental circumstances may sometimes cause the torpidity of individuals of all.' Forster concludes his rambling discourse by saying: 'The result of my researches on this subject has convinced me, that the swallow is a migratory bird, annually revisiting the same countries in common with other birds of passage.'[43]

Hurrah! A firm statement at last. But no, Forster cannot resist adding a caveat: '... while it is pretty certain that the greatest number of swallows migrate, it is not impossible that many individuals of each species may be concealed during winter near their summer haunts'. Forster's was the last little vote for hibernation; within a year or two any mention of hibernating swallows had disappeared completely from bird books.

Table 1. Authors who give torpor or submersion as a possible reason for the disappearance of swallows, versus migration. The turning point seems to have occurred shortly after 1800.

Date	Author	Torpor/Submersion	Migration
1250	Albert the Great	+	+
1358	Conrad von Megenberg		+
1555	Olaus Magnus	+	
1555	Pierre Belon		+
1597	Gerard the Herbalist	+	
1600	Ulisse Aldrovandi	+	
1603	Caspar Schwenckfeld	+	
1651	William Harvey	+	
1660	Jean-Baptiste Faultrier		+
1660s	Thomas Browne		+
1660s	Johannes Hevelius	+	
1678	Francis Willughby and John Ray	+	+
1691	John Ray		+
1702	Baron von Pernau		+
1724	Daniel Defoe		+
1733–63	Johann Leonard Frisch		+
1742	Charles Owen (in Garnett, 1969)	+	+
1742–3	Johann Zorn		+
1743–51	George Edwards		+
1745	Anonymous		+
1747	Mark Catesby		+
1750	Jacob Theodor Klein	+	
1750s	René Antoine de Réaumur	+	
c. 1758	Carl Linnaeus (in Brusewitz, 1979)	+	
1760	Peter Collinson		+
1764	Johan Leche (in Brusewitz, 1979)		+
1768	Thomas Pennant		+
1771	Alb von Haller (in Roger, 1997)	+	

Date	Author	Torpor/Submersion	Migration
1771	Charles Bonnet (in Roger, 1997)		+
1772	Daines Barrington	+	
1774	Oliver Goldsmith	+	+
1775	James Cornish	+	
1779	Comte de Buffon		+
1780	John Legg		+
1789	Gilbert White	+	+
1790	William Smellie		+
1795, 1796	Johann Bechstein	+	+
1797–1804	Thomas Bewick		+
1802	George Montagu		+
1805	Georges Cuvier	+	
1808	Thomas Forster	+	+
1812	Thomas Gough		+
1823	John Blackwall		+
1823	Christian Ludwig Brehm		+
1824	Edward Jenner		+
1829	John Knapp		+
1830	Robert Mudie		+
1832	Sarah Waring		+
1835	Edward Stanley	+	+
1835	James Rennie		+
1836	Frederic Shoberl		+
1837	James Cornish		+
1846	Leonard Jenyns		+
1852	Anne Pratt		+
1859	Francis Buckland		+
1871	James Ward		+

The playwright August Strindberg was still promoting the idea of swallows spending their winters underwater in 1907. I haven't included this in the table because Strindberg was no naturalist.

The answer to Ray's questions about how and why certain birds migrate came not from the study of wild birds as we might expect, but from birds held in captivity. Caged migrants, like nightingales and warblers, very obviously become agitated around the time of their spring and autumn migration. Hopping restlessly back and forth and wing-whirring throughout the night, they are effectively flying on the spot. I think that, despite his passion for birds, or probably because of it, Ray cannot have kept a caged nightingale, for had he done so he surely would have seen this seasonal agitation. What is more surprising is that no one mentioned it to him. Indeed, the earliest comment I can find on 'migratory restlessness' is from 1707, two years after Ray's death, from the unknown author of *Traité du Rossignol* [*A Treatise on the Nightingale*]:

> in the month of February or March, or mid-September, those [nightingales] who are in the room or in the cage, become impatient and resist much then during three or four days of full moon, and fly against the glass or the frame of the cage, the evening, the night and the morning, as if they feel in themselves then something I do not know what which obliges them to leave the place where they are; what they do not do at another time. And it is this instinct and inner guide that makes them fly by a favourable wind directly to the place where they want to go.[44]

With uncommon insight this unknown bird man not only describes the phenomenon, but recognises it for what it is: sublimated migration.

Little by little, and possibly inspired by *Traité du Rossignol*, others started to add their own observations, Buffon among them, who said of the nightingale: 'Everywhere they are known as migratory birds, and this innate habit is so strong in them, that those which are kept in a cage, are very restless in spring and autumn, especially during the night, during the periods normally assigned to their migrations ...' With great accuracy, suggesting he had witnessed it himself, Buffon also described how caged quail became agitated in autumn and spring at exactly the time the wild birds were migrating.[45] The birds became restless and fluttered with:

> unusual agitations regularly at the season of migration, which returns twice annually in September and April. This uneasiness lasted thirty days each time, and began constantly an hour before sunset. The prisoners moved backward and forward from one end of the cage to the other, and darted against the net which covered it, and often with such violence, that they dropped down stunned by the blow. They passed the night in these fruitless struggles ...[46]

In 1797 the great German amateur ornithologist, bird-keeper and professional farmer Johann Andreas Naumann described how the golden oriole, of which he had several in an aviary: 'sings until he leaves us, which happens at night at the end of July or early August. They became always restless when the migratory period started and flew back and forth in my aviary. This lasted until November. From this fact one can conclude that this bird migrates very far, presumably all the way to Africa ... In March they became restless at night again.'[47]

Today, the German word *Zugunruhe* is used by English and German speakers alike to describe this migratory restlessness.[48] Bird-keepers not only saw the link between *Zugunruhe* and the seasonal timing of migration, they were also able to deduce from the diurnal pattern of restlessness whether their captives migrated at night like warblers and quail, or by day like the starling.

In the 1960s, two hundred and fifty years after its first mention, a young German ornithologist, Eberhard (Ebo) Gwinner, based at the Max Planck Institute at Seewiesen in Germany, decided to conduct the first scientific study of *Zugunruhe*. Fascinated by the phenomenon, his idea was that birds use an internal clock – a possibility first suggested in the 1940s – to find their winter quarters. Ever since Aristotle, the migratory urge in birds was assumed to be instinctive.

The nocturnal agitation exhibited by Johann Andreas Naumann's captive golden orioles in the early 1800s enabled him to predict where they spent the winter. (From J. F. Naumann, 1905)

Perceptively, the author of *Traité du Rossignol* had commented: 'One could say that we do not know the real cause of the changing places [i.e. migration], and that God has put into the nature of these birds, as well as in that of the other migratory birds, a certain instinct and a certain inclination, to change place for their own need, without us being able to discover the reason.'[49]

Based on Naumann's observations of golden orioles, Gwinner believed this 'certain instinct or inclination' to be a physiological device that generates enough *Zugunruhe* to carry a migrant to its winter quarters and no further. To test his idea he chose two closely related warbler species; the chiffchaff, a short-distance migrant that winters in southern Europe and North Africa, and the willow warbler, a long-distance migrant that overwinters in central and southern Africa. By today's standards it was a modest experiment, with just a few birds in each group, but the results were clear-cut and as predicted: the willow warbler's *Zugunruhe* lasted much longer than the chiffchaff's.[50] Not everyone was convinced, however, and to counter the criticism that the birds were restless only because they were not in their winter quarters, Gwinner conducted another ingenious experiment that involved putting some warblers on an aeroplane and flying them directly to their wintering areas in Africa to see if their *Zugunruhe* ceased. It didn't, elegantly confirming that local conditions have no influence on restlessness. The link between the duration of *Zugunruhe* and the migratory distance was a tremendous discovery and in 1974 the German Ornithological Society awarded Gwinner the first Erwin Stresemann Prize for his work. I hope he acknowledged Naumann in his acceptance speech.

The award was well deserved, for Gwinner went on to demonstrate how the onset and the duration of migration depended upon an internal clock. Quite where the clock is in the bird's brain has still to be determined, but we know that this clock also regulates other events associated with migration, such as the extraordinary autumnal gluttony that generates the huge layer of body fat migrants use to fuel their journey.[51]

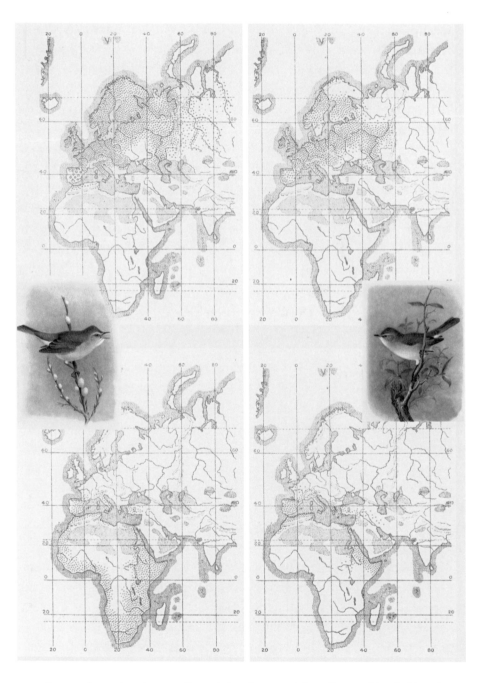

Summer (upper) and winter (lower) distributions of the willow warbler (left) and chiffchaff (right). The willow warbler winters further south in Africa than the chiffchaff, as indicated by these maps from Eliot Howard's British Warblers *(1914).*

For several years following Gwinner's remarkable discoveries it was assumed that internal clocks occurred only in migratory song-birds like warblers and swallows, but a chance encounter with a shorebird showed otherwise. In the 1990s Dutch ornithologist Theunis Piersma discovered a knot that had been living as a family pet for almost twenty years. He had been found on the north Dutch coast in 1980, rendered flightless by a damaged wing. Although fairly fast on his little legs, he was captured, nursed and nurtured in the home of a middle-aged couple, Jaap and Map Brasser, and their black dog Bolletje. Fully grown but of unknown age when he was rescued, Peter was one of millions of knot that spend their winter in the Netherlands after breeding in Greenland's Arctic tundra.

Peter became completely tame, but despite Jaap's physiotherapy never regained his powers of flight. As his owners watched television in the evening Peter stood between their feet, alert during wildlife programmes whose interesting sounds kept him awake, otherwise, asleep. The bird and the dog became close companions and long after the dog had passed away, a tape recording of Bolletje barking would bring Peter running in search of his old friend.

In March 1997, concerned about Peter's legal status and how long he might live, Jaap and Map contacted Piersma, known in the Netherlands from his research as the 'TV knot man'. Visiting them, Piersma was enchanted but also recognised the rather unique research opportunity Peter offered and enlisted Jaap and Map as volunteer research assistants. Every Friday they weighed Peter on a letter balance and noted the colour of his plumage. Piersma wanted to know whether there was an internal clock orchestrating Peter's annual change in weight and plumage, as there is in songbirds. Three years of dutifully collected data revealed unequivocal evidence for a clock: each spring Peter's weight increased from 130 to 190 g, coinciding with the development of his red breeding plumage. When his body weight subsequently declined in the autumn Peter moulted out his red feathers. Prior to this study – the ultimate form of citizen science – annual rhythms were known only for songbirds. Peter clearly had an internal clock, but his timing was idiosyncratic with a cycle spanning eighteen months rather than the usual twelve seen in songbirds. Whether this was due to his age or because he was kept mainly indoors is not clear. Knot are intensely social birds, living in vast flocks during the winter, and it is possible that Peter's ties with his adopted family were more important than his ties to the seasons. As his biographer described it, Peter: '. . . developed his own personal cycle and . . . stayed red as long as possible hoping that Jaap, Map and the dog would also become fat and change colour, after which they would all depart for Greenland.'[52]

In the mid-1700s bird enthusiast Johann Leonard Frisch decided to conduct an ingenious experiment on the swallows breeding near his house in Germany:

A red knot with a damaged wing, kept as a pet by a Dutch couple in the 1980s, provided the first evidence that birds other than passerines have an internal clock regulating their annual cycle, including the switch from red breeding plumage to grey winter plumage. (From Selby 1825–41)

> Because of its fabulous wintering quarters, I have tied threads coloured red with water-colour paint, to the legs of several live-caught swallows, shortly before their departure. The threads were applied as rings, of which the paint would certainly be washed off if it had stayed a short time in water; the swallows however came back to their nests in spring with their red threads on their legs . . .[53]

Through this simple device Frisch demolished the idea of swallows spending their winter under water, but simultaneously confirmed what had been known for centuries for carrier pigeons and suspected in other species: that certain birds have an uncanny ability to find their way 'home' year after year.

Birds had been individually marked on several occasions prior to Frisch's experiment. Pigeons in ancient Greece were 'marked' with the messages they carried, and medieval falconers sometimes placed metal rings of ownership on their birds.

In 1702 another bird enthusiast, Baron von Pernau, proposed cutting off a toe from nightingales and fieldfares in order to identify them and to establish whether they returned to the same area each year. As far we can tell he never actually did this, or if he did his toeless birds didn't return and he didn't write about it.[54] Edward Jenner – best known for his discovery of vaccination but also fascinated by birds – removed claws (only marginally more humane than removing toes) to identify swifts and succeeded in confirming that the same birds breed at the same location year after year. Luckily for Jenner, one of his birds was caught by a cat seven years after he first marked it.[55]

Despite these pioneering experiments, the idea of marking birds as a way of discovering more about their migrations was long thought to be impractical. Finally, in 1899 a Danish schoolteacher, Hans Christian Mortensen, fashioned his own aluminium numbered rings for starlings. Some distant recaptures convinced him and everyone else of the value of this new approach and ornithology was never quite the same again.[56]

Once the use of rings confirmed that birds could find their way, the next question was how they did it.

In a chance observation, the German zoologist Gustav Kramer noticed that the way a bird positioned itself in its cage indicated the direction in which it would migrate if it were free – an observation that changed the direction of migration studies. It is perhaps surprising that this discovery should have been made as recently as the 1940s. People had watched their cage-birds for centuries: why hadn't they noticed the directionality of their migratory restlessness? Once they had noticed it, Kramer and his students constructed special cages to measure it. This was low-tech, labour-intensive science, with observers lying on the ground beneath wire-bottomed cages in order to score which perch the bird was on and the orientation of its body. In later models, perches were fitted with micro-switches and the information was collected automatically, but these so-called Kramer cages still had their limitations – they were bulky and expensive so only a few birds could be monitored at any one time and, if used outside, the damp played havoc with the micro-switches.[57]

Frustrated by these features but keen to monitor large numbers of migrant birds in the field, American biologist Steve Emlen, a doctoral student at the University of Michigan during the 1960s, sought a cheaper, more practical orientation cage:

I began experimenting with funnel-shaped cages constructed out of thin sheets of aluminum and observing the bird's behavior from underneath. I became convinced that many species exhibited Zugunruhe by jumping repeatedly up the sides of the funnel cages, and that this jumping was often directed in the appropriate migratory direction. These first funnel cages were promising: they were small, easy to transport and contained no moisture-sensitive electronic switches. But they still required observation of the bird's behavior inside the funnel.

In the fall of 1963 I travelled from Ann Arbor, Michigan, back to Madison, Wisconsin, to spend Thanksgiving with my parents. While

*The cheap and cheerful orientation funnel invented by Steve Emlen in the 1960s
to document the migratory restlessness of birds like the indigo bunting. The ink pad
in its base cause the birds' feet to leave an inky trace on the sides of the funnel,
indicating the direction of migration.*

there I discussed my successes and (mostly) failures in orientation cage
designs with my father. My father, John T. Emlen, Jr., was a highly
regarded ornithologist and behavioral ecologist at the University of
Wisconsin. In our ensuing discussions, it was my father who first suggested using paper inside the aluminum funnels and letting the bird
leave its own record on the paper. The first suggestion – to use paper
in the funnels – rapidly evolved into using blotter paper. Blotter paper
proved sturdy enough that the paper, alone, became the funnel. No
aluminum backing was needed. The second suggestion – to let the
bird leave its own record – went through several intermediate designs
(scratch papers; carbon paper; crayon over writing paper, etc) before
we settled on using printers' ink and having the ink pad become the
bottom of the new funnel cage design.[58]

The bird picked up the ink on its feet from the pad and left an inky
trace on the blotting paper as it hopped during its *Zugunruhe*. The

density and orientation of the ink marks allowed Emlen to quantify both the amount and direction of the bird's overnight hopping – a wonderfully simple device that revolutionised the study of bird migration by allowing researchers to have several running at a time.

The results of these studies strongly implied that many small birds were genetically programmed to migrate in a particular direction. It sounds straightforward, but to maintain a particular direction a bird still needs a compass. The author of *Traité du Rossignol* speculated in the early 1700s that nightingales might use the moon to guide them on their nocturnal migrations,[59] but there is no evidence for this. They do, however, use the stars. Or at least, nocturnal migrants do. For birds that migrate during daylight, the sun, coupled with a sophisticated diurnal clock that allows them to correct for the sun's movements, serves as their compass. On their own, star and sun compasses are not enough to keep birds on course. Two further cues are important: the earth's magnetic field and the birds' sense of smell.[60] You might ask why birds need so many navigational systems, but they are fail-safes; some work better than others under certain conditions, and natural selection has ensured that if one system fails, there is a back-up.

Aristotle, Frederick II and John Ray all assumed migration to be innate, which today is the same as saying that it has a heritable or genetic component. Some of the most compelling circumstantial evidence for an innate process was the fact that young birds with no experience and no adults to follow managed to find their way to their winter quarters and back. In the 1950s a Dutch ornithologist, Ab Perdeck, decided, in what was to be one of the most ambitious migration experiments ever undertaken, to test the idea that birds were directed to and from their winter quarters by an inbuilt compass,

Since the seventeenth century Dutch bird-catchers had operated *vinkenbaans*, or 'fowling yards', to catch migrating starlings and other small birds for human consumption. Although this kind of bird-catching had ceased in the 1930s, Perdeck had no trouble relocating the men and their equipment to help him catch the large number of birds he needed for his experiment. The study was conducted over several years between the late 1940s and early 1950s, over which time some 11,000 starlings were captured during their autumn southward migration. Each bird was ringed, its sex determined from its plumage and its age (young birds, hatched that year, and adult birds) determined from the development of its skull, which could be seen through the skin by wetting the feathers on the bird's head. Placed in cardboard boxes, the birds were taken on the back of a motorbike to Schiphol airport. They were then flown to Switzerland, released and allowed to continue their migration – except they were now displaced some 600 km (about 380 miles) to the south-east. Perdeck's question was this: would they continue their migration as though nothing had happened and end up 600 km south-east of where they should have been, or would they 'know' where they were supposed to be going, adjust their migration route and end up in the correct wintering area? If the birds were merely using an internal compass, the expectation was that they should continue flying in the correct south-westerly direction, but displaced 600 km to the south. On the other hand, if they ended up in the correct wintering area this would indicate that they knew where they were supposed to go: a true navigational ability.[61]

The key to the success of this experiment was the public. Perdeck relied on them to tell him if they found one of his ringed birds, alive or dead, and to increase the chances of this happening he publicised his study. A grand total of 354 ringed birds was subsequently reported and they revealed something quite extraordinary. Young starlings behaved as though they had an internal compass and ended up south of their normal wintering area, but the adult birds ended up in their regular winter quarters, apparently compensating for their experimental displacement. The obvious explanation for this difference between the young and adult birds was that migration is not based entirely on an innate sense of direction but also relies on experience. For a young bird with no experience, an innate programme dictating both direction and distance is probably the best evolutionary option. Older, more experienced birds seemed to be able to use their experience to adjust their migration, allowing them to compensate if they were blown off course by adverse weather or an ornithologist.

Confirmation that this was the case came from looking at Perdeck's experimental starlings later in their lives. After wintering in the 'wrong' area, the young birds migrated north-east again in the spring and, remarkably, returned to breed exactly where they had been reared themselves – not 600 km south-east of this area. Even though they had been displaced on the first southward migration, they somehow managed to return to their natal area to breed. However, when these birds came to their second autumnal migration they flew directly to the 'wrong' wintering area again! These curious observations had a simple but remarkable explanation. Young starlings learnt something about the area in which they were reared; they 'imprinted' on that area. When they set off on their first southward migration their innate internal compass sent them in a particular direction, and if, as in Perdeck's experiment, they were displaced and with no way of knowing where they should have been going, they simply flew in the genetically programmed direction and spent the winter in the 'wrong' area. In the spring, however, they used both their internal compass *and* their experience – obtained through imprinting – to guide them back to the correct breeding area. When they migrated south again that autumn they returned to the 'wrong' wintering area in which they had spent their first winter, because they had also imprinted on to this.

Perdeck's wonderful study suggested two things about migration. It provided additional support for the idea that the direction and duration of migration is heritable. It also showed that birds imprint on to an area, by developing a mental image of where they need to go, and are then able to make adjustments to the genetic programme if the need arises.

For the last page in this story we return to the Radolfzell castle and meet the director of the Max Planck Institute for bird migration. Slightly portly and sporting an enormous Rip Van Winkle beard, Peter Berthold exudes cheerful Teutonic confidence. His life's goal has been to establish once and for all whether the duration and direction of *Zugunruhe* are heritable. The blackcap is his species of choice, a bird

BLACKCAP
SYLVIA ATRICAPILLA.
251

Crossing European blackcaps (top: female; bottom: male) *from migratory and non-migratory populations and comparing the behaviour of the offspring with that of their parents provided the much-needed evidence for the genetic basis of migration.* (From Dresser's Birds of Europe, 1871–81) common across much of Europe with populations that migrate both long and short distances, and some – like those on the Cape Verde Islands – that don't migrate at all. This is ground-breaking, but also back-breaking research, for the effort required is immense. Berthold is lucky, for the essence of the Max Planck Institute is to support the kind of research no one else can do.

Berthold's first objective is to establish whether blackcaps from different regions, and hence with different migration directions and distances, show different patterns of *Zugunruhe*. One study population consists of locally caught blackcaps, but the other is a partially migrant population from the Canary Islands. The birds for these experiments have to be naïve, with no experience of migration, so the safest way of ensuring this is to take young birds from the nest and hand-rear them. As anyone who has done it will know, hand-rearing young birds is hard work as they require feeding several times each hour from first light until dark. Eventually the chicks are fledged and once their autumnal *Zugunruhe* starts they are placed in the orientation cages. As expected, the birds from two populations show quite different durations and directions of restlessness, giving Berthold the green light to start planning the real experiment.

His plan is to hybridise blackcaps from the two populations and see how their offspring perform in the orientation cages. It sounds easy: take one bird from each population, pair them and wait for the babies to appear. Left to their own devices in an aviary the birds sing, copulate, lay eggs and hatch their chicks, but they are incapable of rearing them. Without their specialised insect diet, available only in the wild, the blackcaps cannot raise their chicks. But Berthold finds a solution by enlisting a team of field assistants who scour the local woods for dozens, sometimes hundreds, of wild blackcap nests. Shortly before the captive birds' eggs are due to hatch they are swapped with eggs from a wild birds' nest (at an earlier stage so they don't hatch in the aviaries). The wild parents rear the aviary chicks until they are a week old, at which point the researchers transfer the grown nestlings back into the aviary. I have only ever found a couple of blackcap nests (admittedly without looking very hard), but the effort needed to find the nests, check their status and swap the eggs and chicks was enormous. But it is worth it. The results are fabulous – the hybrid offspring exhibit a pattern of *Zugunruhe* intermediate between either of their parents' populations. Their orientation is intermediate; the duration of their restlessness is intermediate, providing compelling evidence that *Zugunruhe* is heritable.[62]

I was a young lecturer when these experiments were completed and the results were first aired in public at the International Ornithological Congress in Moscow in 1982. I remember my Ph.D. supervisor Chris Perrins, who was not normally effusive, coming back from the meeting in a state of great excitement and recounting how Berthold's work had stolen the show, describing it as one of the most remarkable bits of ornithological research ever.

Peter Berthold's results also helped to explain a new migratory pattern in a wild population of blackcaps. During the 1970s birdwatchers were starting to find more and more blackcaps wintering in Britain rather than migrating to Africa as they had done previously. A combination of extensive winter bird-feeding and climate change now allows these blackcaps to overwinter successfully in Britain. When he checks, Peter finds that a *tiny* proportion of blackcaps caught in Germany exhibit a tendency to migrate in a north-westerly direction (instead of southwards), but of those wintering in Britain the *majority* show the same thing. The combination of more food and milder temperatures in Britain means that the birds that migrate from Germany to Britain are now apparently at an advantage over those migrating south. Spared the long flight to and from Africa, their overwinter survival rate may be relatively high, and they may also gain a second advantage by returning to the breeding grounds and securing a territory earlier in the spring than the birds that migrate to Africa. Prior to the environmental changes in Britain, the few blackcaps that set off in a north-westerly direction probably paid a high price for their inheritance.[63]

One of the most popular of cage birds, captive nightingales observed in the 1600s and 1700s provided some of the first clues to the biological bases of migration. (From H. L. Meyer, Illustrations of British Birds, 1835–50)

5

Illuminating Discoveries

Light and the Breeding Cycle

The nightingale is the ultimate songbird. Its exquisite song has moved men to tears, sent poets into raptures and driven bird-trappers to the woods to capture it. With its deep, almost sobbing quality and poignant pauses, there is indeed something extraordinarily evocative about the song of the nightingale. The desire to claim ownership of this song meant that for centuries nightingales were trapped and incarcerated in huge numbers. As a cage-bird, however, the nightingale was but an ephemeral delight, for most individuals expired within a day or two and those that did survive sang for only a few weeks each spring.

By comparison, the canary was so much easier. It was tame, bred readily in captivity and sang throughout much of the year. The canary's song was loud, cheerful and varied, but not a patch on that of the nightingale. To get the best of both worlds bird-keepers since the Middle Ages had fantasised about hybridising the two species, but despite some bold attempts were never successful.

But remarkably, an attempt to combine the nightingale's song with the canary's robust constitution *did* work. In the 1920s a German bird-breeder by the name of Karl Reich created a strain of canaries that sang like nightingales. Their song was so perfect that it fooled ornithologists into thinking Reich had real nightingales in his apartment. The birds won prizes; Reich's recordings of their song sold across the world and everyone wanted to know his secret. Envious of his success, some even accused him of dishonesty, but Reich's only dishonesty was in not telling competitors how he created these extraordinary songsters.

Bird-breeders knew full well that young canaries could be trained to sing almost anything they heard, be it another canary, a linnet or even a penny whistle. What they never heard, however, was a canary singing a nightingale song because the nightingale's singing season was so short that they had ceased singing by the time young canaries had left the nest and were ready to learn a song.

Ingeniously, Reich discovered a way of persuading nightingales to sing at exactly the time his young canaries were learning what to sing. How had he done it? The trick was to shift the nightingales' annual cycle so that they started singing later than usual and continued to sing well into the summer. To put a stop to the questions, Reich told people he had discovered the method in an old book (which was true) and, rather less honestly, that it involved raising the temperature at which he kept the birds over winter. It sounded plausible, and, remarkably, no one ever bothered to check. To be fair this would have been tricky; the book was very old – it was published in 1772 – and very rare, but had they done so all they would have found was a description of the bird-catchers' ancient practice of 'stopping' birds.[1]

'Stopping' in this context meant stopping the light. What Reich failed to tell his customers and competitors was that he delayed his nightingale's song by altering the amount of light they received. It was an old trick, but one that proved to be the basis for understanding how birds organise their annual cycle.

A bird's year is a succession of events, including breeding, moulting (the annual replacement of feathers), and, for certain species, migration. The timing of each of these within the annual cycle is crucial. Getting it wrong can spell disaster. In temperate regions birds typically breed in the spring, and migrate south in the autumn and north again the following spring, with moult taking place either before or after migration – and occasionally at the same time. How do birds know it is spring and time to breed? How do they know when to start dropping feathers and growing new ones, and how do they know when to set off on their migrations? Resolving how birds time the different events in their annual calendar has been one of the major success stories in ornithology.

For centuries bird-trappers had captured small birds on their southward autumnal migration. The oldest and most effective method for catching birds was the clap net. It had been used by the ancient Egyptians and was subsequently described in many of the early bird books, including Ray's and Willughby's *Ornithology*.[2] On their own, however, clap nets were no good. Decoy birds, or 'stales' as they were sometimes called, were essential if the nets were to be effective. There were two types of decoys. First there was the 'flurr bird' (also known as a brace bird), located in the centre of the catching area. The unfortunate bird was attached to a swivel and a short line by means of a harness made from soft leather, string or silk, at one end, and a hinged stick operated by means of a long line at the other. As migrating birds flew overhead, the trapper pulled the flurr stick, flipping the decoy into the air and then, as the tension on the string was released, the harnessed bird fluttered downwards as though it was alighting. The other type of decoy bird was housed individually in small cages around the perimeter of the net; their job was to sing. Migration and singing usually occur at different times of year, so arranging for decoy birds to be singing at the time when their wild counterparts were migrating was a problem. Today, bird-ringers (banders) have appropriated and made respectable many of the old bird-trappers' methods, and routinely use recordings of birdsong to lure migrants into their nets.

One of the earliest accounts of stopping[3] comes from the Italian bird-keeper Cesare Manzini in 1575, but the technique may well have been used much earlier. Here's how he describes the process:

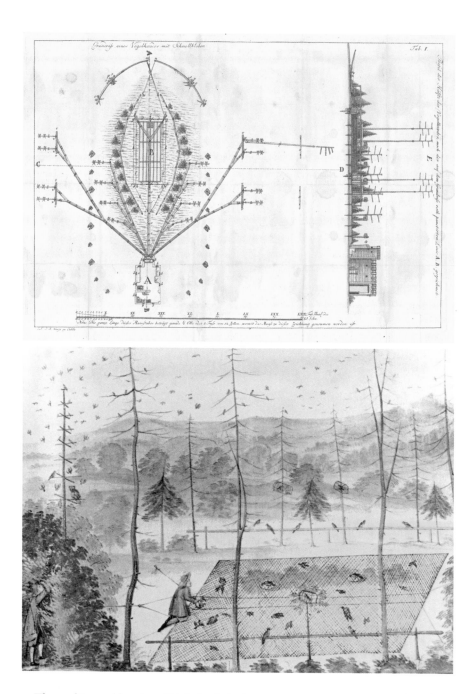

The catching and keeping of birds for their plumage or song was once extremely wide-spread in Europe. Trappers employed large clap nets and decoy birds in full song to lure the wild birds into the nets. (From Bechstein (top), 1801–22, and Birkner, 1639)

Although all other birds, with the exception of the Chaffinch, sing in winter, i.e. the Goldfinches, Linnets, Greenfinches and similar species, there are nevertheless others, which after having come out of their partition, cease to sing because of the moult. Therefore from the beginning of May, you purge those which will be used in similar fowling business, as following, first you give them beet sugar, mixed with a little pure water; the following day a leaf of the same herb; the third following day, put him on the ground, so that they eat during a time span of ten days, and close it off each day a little from the light in the dark. After ten days, you give it again beet, putting it in a box in a dark and isolated place. In the evening you manage by the light of a lantern, which is seen by the birds during two hours; and in that time you can clean its drinking-trough, and every eight days change the hempseed and give also every four days beet leaf, and sugar every 20 days to them, especially to the Chaffinch, which easily become blind. In order to prevent them from having poxes, it is necessary that every 20 days the cage is changed, for an additional reason, i.e. the stench which can easily kill them. These things have to be done until the 10th of August, from that time on, you have to purge them again, in the same way as above, allowing them gradually to see the air, until the twentieth of the same month, taking care not to keep them in the sun. So they will serve well for hunting and fowling in the months of September and October . . .[4]

As subsequent accounts make clear, the 'purging' with beet leaves wasn't essential. The key element was reducing and finally 'stopping' the light, which had the effect of accelerating the birds' annual cycle and rushing them through an artificial autumn (and moult). Then, after three months in the dark, the light was gradually increased so that the birds emerged in reproductive condition to a 'new spring' that was in fact autumn.

Reich employed a modification of this method, keeping his nightingales at low light levels for an artificially extended winter and by so doing delayed their moult so that they started to sing two months later

Overleaf: Observations of birds in captivity contributed enormously to the development of ornithology. This image by Emil Schmidt, showing a selection of foreign birds (left) and European species (right), is from Karl Russ's volume on bird-keeping published in 1888.

than normal. In fact, Reich excelled at manipulating his birds' annual cycle and was so proficient that he could arrange for nightingales to be singing at any time of the year.[5]

Singing decoys were essential whatever trapping method was employed and almost every bird-keeping book, from Manzini's onwards provides an account of stopping.[6] Some of these descriptions include elaborate procedures, but even a relatively modest reduction in light seemed to have had the desired effect. Luckily for Reich the practice of stopping had all but died out by the early twentieth century, which is why he was able to keep his method secret.

The bird-keepers' 'stopping' trick was very slow to filter through to the ornithological community, and by the time it did they had discovered the effect of light for themselves. More unexpectedly, the effect of stopping on a bird's breeding cycle seems to have been just as slow to infiltrate other areas of bird-keeping. Canary-breeders would have loved to have had birds in full song throughout the year, and especially in winter, but somehow never seem to have made the connection between stopping and singing. Had they done so, they could have had some birds singing through the dark winter months. They came close, however: for some canary-breeders in the late 1700s knew that by means of artificial light and heat they could get their birds to breed outside the normal breeding season:

> It appears to be not sufficient that one pleases oneself in summer with the breeding of canaries. And so I am often asked whether canaries would breed during the winter. Not only is this possible, it also succeeds. Moreover it is very amusing for the bird fancier who is active at home in winter. But I have to say that not all canaries are capable, and, if one does not have a warm room, clearly lit, everything is done in vain.[7]

Canary-breeders never identified light as the cue; instead, as their accounts intimate, they considered temperature to be more important. Poultry-breeders were exactly the same. The great French naturalist-cum-polymath René Antoine de Réaumur was commissioned in the mid-1700s to industrialise French poultry-breeding and, after heating their cages, fully expected his chickens to start laying in midwinter. They didn't, and he was forced to conclude that it was their moult rather than the cold that stopped them laying. He also recognised what we now know to be an energetic trade-off between moulting and laying:

> all the time it [the moult] lasts, the wasting of the nutritive juice which is made by and for the unfolding and increase of the new feathers is considerable, and it is no wonder that there should not remain enough of it at that time within the body of the hen to cause eggs to grow within her.[8]

Réaumur then speculates that, were it possible to make a hen moult earlier in the year, she would duly lay eggs throughout the winter. He even fantasises about plucking all the feathers from hens to achieve this, urging his readers to try the experiment for themselves. Réaumur also cautions that they should not remove the feathers all at once, but 'nature is to be imitated by taking them from her gradually'. It is interesting that, for all his ingenuity, Réaumur was apparently unaware of the bird fanciers' practice of 'stopping' to force a premature moult and the role that light played in regulating its timing.

One of the first explicit statements that light was important to birds came from neither a bird-keeper nor an ornithologist, but from a Finnish poet, Johan Runeberg, who, watching birds from his sick bed in 1870, wrote a poem called 'The Lark' in which he says:

> Only the sun it will follow, over land and ocean,
> All the way to the southern home and to the north.[9]

It is not clear when Runeberg wrote this particular poem, but his insight was brought to the attention of ornithologists in an anonymous article published in 1874 that stimulated a flurry of letters to *The Times*, including one from the English ornithologist Alfred Newton, in an attempt to squash Runeberg's idea. It is not hard to see why Newton was so negative. The suggestion that birds migrated in search of longer days, much as they might search for more food in milder climates, simply did not stand up to scrutiny. As Newton pointed out, some migrants set off before the autumn equinox, and by doing so move into areas with shorter, not longer days. Newton was right, of course, but he missed a glorious opportunity. Had he stood on Runeberg's shoulders rather than on his idea, he might have revolutionised this area of ornithology.

Runeberg's 'idea', if we can call it that, was that light was the evolutionary cause of migration, rather than the cue that initiated it. There is a subtle but extremely important difference between these two points, and one we shall come back to later.

Some fourteen years after Newton had successfully demolished Runeberg's idea, a well-known amateur ornithologist and successful Sheffield steel manufacturer, Henry Seebohm, made an almost identical suggestion regarding the migration of wading birds.[10] It is unclear whether Seebohm was aware either of Runeberg's poem or Newton's letter, but it seems unlikely, and it was irrelevant, for no one noticed his perceptive idea. In the early 1900s the link between light, migration and evolution got yet another airing when the physiologist Edward Schäfer mentioned it at a meeting of the Edinburgh Natural History Society in a talk entitled 'On the incidence of daylight as a determining factor in bird migration'. Born in Hamburg but educated at University College London, Schäfer was later knighted for his physiological research and in 1918, in a somewhat eccentric move, changed his name to Sharpey-Schäfer, in honour of his friend William Sharpey, the father of English physiology. It isn't obvious that Schäfer had any special interest in birds, but he was fascinated by hormones and the way they might influence the seasonal activities of birds. Schäfer's main theme was based on the ideas of both Runeberg and Seebohm: birds were in search of daylight rather than food. By flying south in autumn they sought longer days, just as they did when they moved to

Attempts to get canaries to breed in captivity during winter failed because breeders focused on temperature rather than light. These images of different canary breeds are from the Reverend Francis Smith's popular book of 1868.

higher latitudes in the spring. What was novel, and what Schäfer was probably unaware of, was his throw-away comment that day length might be the *cue* for the onset of migration:

> the incidence of the proportion of light to darkness is a constant factor, and might even be conceived to be operative in exciting the migratory instinct into activity.[11]

Schäfer's flash of insight was too brief and too well hidden to be noticed and another thirteen years passed before ornithologists began to wake up to the idea that day length might be the seasonal cue that initiates both breeding and migration. Curiously, the wake-up call was an article not on birds, but on plants, describing how day length (or photoperiod, as the authors called it) provided the environmental signal triggering the onset of flowering. Subjecting plants to different day lengths, United States government botanists Wight Garner and Harry Allard showed in 1922 that they could induce flowering at almost any time of year – in exactly the same way and at almost the same time that Reich was altering his nightingales' singing seasons.

Garner's and Allard's botanical study finally jolted the ornithologists out of their intellectual torpor and got them thinking about seasonal cues. But frustrated by their lack of initiative, Allard wrote his own article on bird migration in 1928, obviously rather thrilled that it was a poet and two botanists rather than an ornithologist who first identified the importance of light. He wrote:

> There is something beautiful in the suggestion, beautiful because science with its cold, modern viewpoints and methods scorning the glamour of poetry, may yet prove the poet's simple inspiration true in greater or less degree.[12]

Today, the acknowledged pioneer of day-length and breeding-cycle studies is the Canadian physiologist Bill Rowan. An eccentric, Rowan worked in glorious isolation, apparently oblivious of what other researchers or bird-keepers might have known. With an interest in migration that reached back to his childhood, Rowan's goal as a professional biologist in the 1920s was to identify the external stimulus for bird migration. Through a series of logical steps he developed the hypothesis that light was the trigger, eliminating temperature and barometric pressure, which were previously thought to be the most likely cues, and recognising that day length was the *only* factor that followed a predictable pattern each year. In her biography of Rowan, Marianne Gosztonyi Ainley writes: 'Daylight, as such, had already been considered by others as important in plant growth and bird migration. But Rowan had no access to, or time to read the work of others, and came to the same conclusions independently.' For several years the overworked Rowan had been content to accumulate data and ponder the problem rather than rush into print. But then, in the summer of 1924, he was roused into action by the appearance of a paper in the American bird journal *The Auk* by Gustave Elfrig. The title of Elfrig's paper asked rhetorically whether photoperiodism – that is day length – was a factor initiating the migration of birds. His hypothesis, stimulated by Garner's and Allard's earlier studies of the effect of day length on the timing of flowering in plants, was a physiological one and he proposed that the spring migratory urge in birds was caused by the growth of their gonads.

Convinced Elfrig was wrong, Rowan was incensed. His own results suggested that day length independently stimulated both gonad development *and* the urge to migrate, rather than the one causing the other. He was absolutely right and later research revealed that birds whose gonads had been surgically removed continued to exhibit migratory restlessness. As so often happens in science (even back in the 1920s), it was competition and the idea that Elfrig might claim the glory that motivated Rowan in the autumn of 1924 to begin his definitive experiments. It wasn't easy. The university president disapproved of Rowan's work; his teaching duties were onerous; funds were limited and he had trouble catching (and keeping) enough birds for his

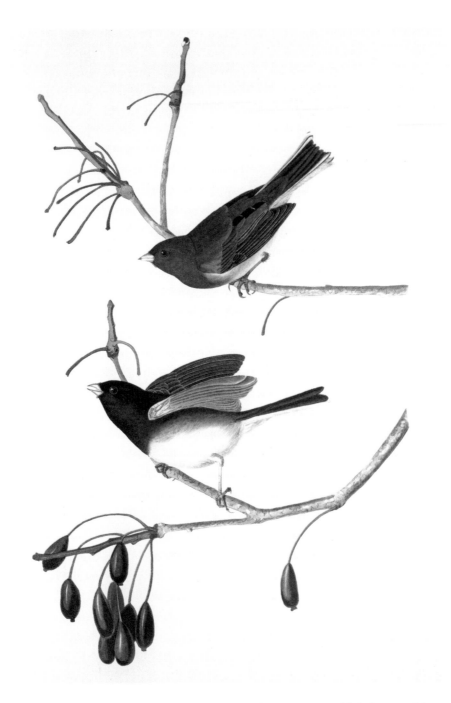

The junco – a North American sparrow used by Rowan to establish the powerful effect of day length on birds' breeding cycles. (From Audubon, 1827–38)

experiments. Rowan persisted, building aviaries in his backyard and fantasising that by artificially increasing day length he might be able to stimulate spring activity in autumn gonads, and 'thereby induce the owners, when released to go north, instead of south in the autumn'. Despite all the obstacles Rowan got his research going and with the extra light provided only by a 75-watt bulb the effect on his birds was spectacular. He wrote ecstatically to a fellow ornithologist:

> I have succeeded in experimentally inducing Juncos [a North American sparrow] to develop spring fever at Christmas in large aviaries in the garden with temperatures running down to 52 [°F] below zero. They were singing all day long ... and on dissection proved to have large spring testicles ...[13]

Rowan never succeeded in demonstrating a reversal of the direction of migration, mainly because it wasn't possible to establish accurately the direction in which his experimental birds set off on being released: the birds simply disappeared into the undergrowth. But in the late 1960s, using Indigo buntings in his special orientation cages, Steve Emlen, then at the University of Michigan, Ann Arbor, elegantly confirmed Rowan's hypothesis by showing that, regardless of the time of year, birds manipulated by artificial light into spring condition hopped in a northwards direction, while those in autumnal condition hopped southwards.[14]

If we believe that Rowan really did not look at the literature and made his discoveries independently of anyone else, it should come as no surprise that he hadn't heard of 'stopping'. With one notable exception, few ornithologists had. In the Netherlands, however, bird-catching was still part of everyday life in the 1930s and a young Dutch researcher named Piet Damsté decided to build on Rowan's results and discover how stopping worked.

Using captive greenfinches, Damsté found that if he simply followed what bird-keepers did and reduced the amount of light available in May, the birds' gonads rapidly regressed, sperm production ceased

In the 1960s Steve Emlen used indigo buntings to test the idea that birds given additional light to bring them into 'spring condition' orientated themselves in his special cages in a northward direction while those manipulated into 'autumn condition' orientated themselves southward. The painting is by Mark Catesby (1741–3).

and they stopped singing. Kept in total darkness, the birds underwent a full moult, dropping their feathers and growing a new set. Then, as Damsté gradually increased the light in August, their gonads rapidly regrew and within three weeks the males, at least, were in full breeding condition, producing sperm and singing all day. For some reason, female birds never quite came into full reproductive condition or laid eggs under this experimental regime, presumably because some crucial factor was missing. Damsté's experiments nevertheless elegantly exposed the underlying processes that bird-keepers (without really caring how it happened) had been happily exploiting for centuries.[15]

Light was clearly the trigger, but how did it work? The answer was complicated and involved a suite of additional questions, the first of which asked how birds actually perceive light. The obvious answer is through their eyes, but in fact it isn't the case. Through some rather grisly – and by today's standards unacceptable – experiments the French researcher Jacques Benoit revealed in the 1930s that birds perceive light through their skull. He showed this first by demonstrating that ducks whose optic nerves had been cut or whose eyes had been surgically removed still came into reproductive condition in response to increasing light levels in the spring. Second, by placing half a (dead) duck's head on photographic paper, Benoit demonstrated that light – and especially red light – penetrates the skin and bone of the skull to reach the brain. This was a totally unexpected and remarkable finding, but you can easily see for yourself how it might work simply by holding your hand over a flashlight: flesh and bone are not impermeable to light. Years later, using fibre optics to light up particular parts of the birds' brain, it now appears that the receptors that detect changes in day length are located in a region of the brain referred to as the basal hypothalamus – the same area that in humans controls body temperature, thirst and our own daily and annual rhythms.[16]

Benoit conducted another experiment that had a huge influence on our understanding of how birds' breeding cycles are regulated. He found, to his surprise, that when male ducks were kept in *total* darkness their gonads continued to show the typical pattern of seasonal growth and regression – increasing in spring and decreasing in size in the autumn. This remarkable observation eventually led to the discovery of internal clocks. Benoit's results revealed that the processes controlling the birds' breeding cycle were more complex than anyone had previously imagined. As well as the seasonal changes in day length – an extrinsic factor – there was also an internal clock – an intrinsic factor – in operation, and the two worked in concert. This was a major discovery, but because of its nature was better known among physiologists and the medical profession than it was to ornithologists. A professor at the Sorbonne in Paris, Benoit was among the academic elite. Confident and flamboyant, he enjoyed fast cars and exploited his medic's badge to speed with impunity.[17] His research should have won him awards, but for some reason the French academic establishment chose not to honour him. I was puzzled, and when I asked my ornithological colleagues why this should be, no one seemed to know.

I eventually discovered an extraordinary series of events that must represent one of the most unfortunate episodes in the history of ornithology. The story begins in the late 1950s with the appearance of a new researcher in Benoit's laboratory. From the start there was something different about *le père* Leroy. Apart from being a Jesuit, he seemed to have had difficulty mixing with other members of the research group. Assuming this was a result of having spent a long period in China, Benoit was sympathetic and gave Leroy a project outside the lab's main line of research and allowed him to work alone. The project Benoit allocated him was topical but rather speculative and involved injecting the DNA of one species of duck into another in the hope that it might induce mutations that would then be transmitted to their offspring. It was a long shot, but if it worked it would fly in the face of conventional patterns of inheritance. Despite Watson's and Crick's discovery of the structure of DNA just a few years earlier, in 1953, belief in the inheritance of acquired characteristics (that is, inheritance via non-Mendelian processes) was not quite dead – hence the duck project.

In the 1940s a Dutch researcher employed European greenfinches to work out how bird catchers used 'stopping' to produce a bird in full song in the autumn instead of the spring. (Painting by Johann Walther from c. 1650)

Leroy tackled the research with enthusiasm and, to Benoit's utter amazement, subsequently announced that the experiments had been successful: the offspring of the injected birds had black or pink beaks rather than yellow ones like their parents! The protocol concerning the revelation of results of major scientific importance was to send a sealed note to the French Academy of Sciences to ensure priority should anyone else claim the discovery for themselves. Hardly able to contain his excitement, Benoit wrote an account of Leroy's findings and sent it off to the Academy.

Not long afterwards, Leroy came to Benoit and showed him a newspaper article in which some researchers in North America had apparently obtained similar results, transforming the colour of a bird by injecting it with the DNA of another species. Benoit was staggered and, keen to secure his priority, felt he had no option but to reveal the contents of his sealed letter to the Academy. Such events do not go unnoticed, and the announcement was made in the full glare of the country's media. Overnight, Benoit was catapulted to scientific stardom. Everyone was excited, including President de Gaulle, who no doubt saw a Nobel Prize on the horizon.

Only after he had declared his hand did Benoit manage to contact the author of the American study, to learn that there had been no such discovery. Benoit was perturbed and puzzled, to say the least. But the other researchers in Benoit's lab knew exactly what had happened. Leroy had set the whole thing up. Not only had he fudged the results and written the newspaper article that pushed Benoit into disclosing his results, he had also arranged for the media to be present when he did so. Benoit refused to believe it, unable to accept that a Jesuit could behave dishonestly and – to his cost – defended Leroy. Well and truly duped, Benoit paid the price. The scientific community never forgave him.[18]

The studies of Rowan, Benoit and others stimulated a huge amount of research on the timing of birds' breeding seasons. It was obvious that in temperate regions like Britain and North America birds reproduced in the spring as day length increased. It was also clear, as all bird-catchers knew, that not all birds reproduced at the same time, suggesting that different species responded to changing light levels in different ways.

Rowan's studies showed that light was an important trigger both for the onset of migration and reproduction. Benoit demonstrated how light penetrated the brain and, in doing so, stimulated the growth of the gonads, which in turn triggered the onset of song in males. The question now focused on the link between the brain and the gonads. The answer was hormones. Chemical messengers had been suspected for a long time, notably by the German physiologist Johannes Müller, and their existence was confirmed by Arnold Berthold in the 1840s when he surgically transplanted the testicles from one cockerel to another. Because Berthold transplanted no nerve cells or other nervous tissue, the experiment strongly suggested that the effect was a chemical one.[19] In a similar vein in the late 1800s the French physiologist Charles Edouard Brown-Séquard injected extracts from the genital glands of monkeys into his patients, with remarkable rejuvenating effects. His widely reported results stimulated both his patients and other researchers.[20]

By the 1950s the foundations of avian endocrinology were well established: light was important; perceived directly by the brain it triggered the release of hormones (gonadotrophins) from the pituitary gland to stimulate the seasonal growth of the reproductive organs.

New hormones continued to be discovered throughout the 1950s and 1960s but the technology to establish *how* they worked remained extremely primitive. It involved killing birds in a particular state – breeding or non-breeding, for example – and attempting to measure the hormone content of their endocrine glands. Accuracy was poor, and it was impossible to measure what researchers really wanted to know, which was the *rate* at which particular hormones were produced.

Hormones were hot and research groups around the world competed to discover how they worked, for whoever succeeded would be assured lasting fame. One team, directed by Jock Marshall and comprising Brian Lofts and Ron Murton, was based at St Bartholomew's Hospital in London. Marshall's interest in this area of research was the result of completing his D.Phil. at Oxford in 1947 under the supervision of John Baker, the pioneer of bird-breeding season studies. A colourful character, Marshall shot all his own specimens despite having lost an arm in a shooting accident at the age of fifteen. He continued his research during the Second World War while serving in North Africa, preserving the gonads of the birds he shot in neat gin, later presenting them to Lofts as the material for his Ph.D. Lofts never used them, however; instead they remained in a jar on his desk as a memento of his brilliant supervisor.[21]

Other endocrine research groups included two in the United States, one led by Albert Wolfson at Northwestern University, Chicago, and another directed by Don Farner at Washington State University. Brian Follett, who became a leading figure in endocrine research, told me how he met the remarkable and energetic Don Farner at a conference in Bristol in 1962. At that time Follett was interested in the physiology of mammals, but was fascinated to learn from Farner that, just like flicking a switch, reproduction in birds could simply be turned on and off with light. Follett went to work with Farner only to discover that he had to go into the field to catch his own research animals – white-crowned sparrows. The focus was entirely on males, not because of sexism, but simply because males came into reproductive condition in captivity, whereas females would not. Follett's job was to pursue the pituitary hormones through the birds' breeding cycle. This was immensely difficult; the amounts of hormone in the blood were almost unbelievably small, and even with 'great numbers of pooled pituitaries' they still achieved only a crude estimate with what he called a 'ghastly degree of precision'. Later Follett switched to Japanese quail which (unlike domestic chickens) were acutely sensitive to photoperiod and could be reared in large numbers and consequently proved a much better species for endocrine research.

As the work developed Follett and his colleagues continued to be frustrated by the poor precision of their measurements. Then, in the late 1960s, hope dawned on the horizon with the development of a new technique known as 'radioimmunoassay'. This method, which involved labelling substances like hormones with radioactive iodine that *could* be measured, was brought to fruition by two American researchers, Sol Berson and Rosalyn Yalow, who won the Nobel Prize in 1977 for using the technique to measure human insulin levels. Follett and his colleagues Colin Scanes and Frank Cunningham were only weeks behind in this discovery:

> On a magical day, yes truly magical, in 1970 we obtained the first results and in one step had leapt across a divide. I exaggerate not. Up to that moment all we could do was kill a bird, take out its pituitary gland and measure its content with an accuracy of plus or minus 100%. The sensitivity of the assay was such that we needed 4000 nanograms[*] of hormone to gain a single estimate. If the changes were large enough in a series of birds then we could speculate upon changes in secretion. The radioimmunoassay would measure 10 picograms[†] and so our gain in sensitivity was 400,000 times. That meant we could take blood samples of 100 microlitres[‡] from a bird and obtain repeat estimates on luetinising hormone content … It is often said that in science technological breakthroughs are the thing – we know what we want to measure but simply do not have the technology – and this proved to be the case for us and many others with radioimmunoassay, it transformed endocrinology.[22]

The team used this new methodology to good effect and, in what is now considered a classic experiment, Follett, and colleagues Philip Mattocks and Don Farner, showed in 1974 that birds had an internal rhythm of sensitivity to light and only when day length *and* the internal rhythm coincided were hormones secreted.

---

[*] 1 nanogram = a thousand millionth of a gram.
[†] 1 picogram = a millionth, millionth of a gram.
[‡] 1 microlitre is a millionth of a litre; 100 microlitres is a few drops.

The timing of different birds' breeding seasons became known only in the nineteenth and twentieth centuries, and were better known by egg collectors than they were by ornithologists. These are the eggs of different European warblers: (from top row) aquatic, sedge, reed, marsh and (bottom two rows) great reed warbler. (From Howard, 1907)

By the mid-1970s the mystery of how birds perceive and respond to light had largely been solved. The remaining question was why events like breeding, moulting and migration need to be timed so precisely. Oxford zoologist John Baker had asked this very question almost forty years earlier, recognising that there were actually two rather different types of question to be addressed if one was to understand birds' breeding cycles. The first, which, as we have just seen, was answered by the endocrinologists, is how birds 'know' when to breed. The second asks what the purpose or benefit of breeding at a particular time of year is.[23]

Baker saw a broad generality in these two types of questions – how and why – for they were relevant not just to birds' breeding seasons but to many other aspects of biology. The first question is about mechanisms; the second is about adaptive significance. Baker called these *proximate* and *ultimate* factors, respectively. The poet Johan Runeberg had confused the two and mistakenly proposed light as an *ultimate* factor in bird migration. By the 1930s it was clear that light was a *proximate* factor; that is, light is the environmental cue that triggers the onset of both breeding and (later in the season) migration.

In contrast, in the 1930s there was little interest in the evolutionary (or ultimate) reasons for the timing of migration or breeding. This is curious for John Ray had discussed this very question in *The Wisdom of God*, writing: 'The ... time of year ... of the production of animals, is when there is proper food and entertainment ready for them.' His ideas had been anticipated by Frederick II more than three centuries earlier, who, in his treatise on falconry, had written sagely that most birds breed in spring because:

This season has, as a rule, an even temperature, which induces an abundance of blood and sperm, and an excess of these two humours arouses a desire in both sexes to indulge in coitus, resulting in racial reproduction. Furthermore, spring is followed by summer, a more favourable season than any other for rearing the fledglings ... If the objection be raised that autumn, owing to its even climate, would be equally propitious for nesting, we reply that if birds were to nest and breed at that season severe winter weather would damage the nest and injure the fledglings before they were mature and feathered and had become accustomed to the cold.[24]

Frederick II and John Ray reached their conclusions about birds' breeding seasons independently and up until the twentieth century were the only individuals seriously to consider what we would now call the adaptive significance of breeding seasons. Ray, of course, couched his explanation in terms of God's providence, but it was none the less still a question about adaptation.

Baker's two types of question had implications well beyond birds' breeding seasons, and in fact became a major concept in biology which was particularly important in shaping and defining the study of animal behaviour. It also divided researchers into those interested in physiology (proximate factors) and those fascinated by ecology, behaviour and evolution (ultimate factors). As a result, for a long time these two areas of ornithological research developed side by side with precious little exchange of ideas. Physiological studies of birds often involved keeping them in captivity and subjecting them to experiments, whereas the ecology, behaviour and evolution of birds meant watching them – with the occasional experiment – in their natural environment. As a young biologist in the 1960s and 1970s I disliked the idea of physiological research, and much preferred watching birds in their natural environment. However, as ornithology and related areas like behavioural ecology matured during the 1990s the value of integrating studies of ultimate and proximate factors became increasingly obvious.[25]

Inspired by the beginnings of a resurgence of Darwin's ideas Baker was ahead of his time in the 1930s, when he said: 'Few subjects call more urgently for an evolutionary and ecological outlook than that of breeding seasons.'[26] Not everyone agreed and Baker's ground-breaking paper on the topic submitted to the premier ornithological journal, *The Ibis*, was summarily rejected by the editor C. B. Ticehurst, who

As early as the 1600s John Ray assumed – correctly – that birds timed their breeding so they had young in the nest when food was most abundant for them. These are young golden orioles. (From Frisch, 1743–63)

was renowned for his conservative attitude. David Lack wrote to vent his frustration about this to Ernst Mayr in the United States, who replied: 'There are a number of good young men in England who could change all this radically, but they never seem to have made an attempt to "Oust the Old Guard". If you can round up sufficient members … you will not have any trouble instituting a new deal.'[27] With this kind of encouragement from Mayr, it is little wonder that Lack later instituted a *major* new deal.

Baker's paper duly appeared as a chapter in a book celebrating the career of Oxford zoologist E. S. Goodrich in 1938. Now, of course, it seems self-evident that birds reproduce at the best time for rearing young, as Frederick II and Ray suggested, but it is easy to be wise with hindsight and many ideas seem obvious once they are articulated. Although Baker could probably not have imagined it at the time, his ideas initiated one of the great ornithological endeavours of the twentieth century. On the one hand he inspired physiologists like Brian Follett to continue their search for the proximate factors that triggered the start (and end) of breeding; on the other he motivated ecologists to establish whether breeding seasons were adaptive. Foremost among the ecologists was David Lack. Ray's hypothesis became one of Lack's most important ideas, although as far as I can tell Lack seems to have been completely unaware of Ray's comment on birds' breeding seasons, which is surprising since Lack was extraordinarily well read.

The key to understanding the evolution of birds' breeding seasons was the fact that different species reproduce at different times of year. The ancient habit of taking young birds from the nest and rearing them by hand meant that birds' breeding seasons – in Europe at least – were known at least approximately from the earliest times. Indeed, the spring breeding of most animals in Europe was so obvious that naturalists like Belon, Gessner and Aldrovandi scarcely bothered to mention it in their encyclopaedias. In contrast, for those intent on acquiring cage-birds or falcons, knowing when different species reproduced was crucial if they were to find nests at the appropriate stage. The nightingale's breeding cycle has been well known since Aristotle's day, in part because males announce their return from migration with a conspicuous vocal flourish, and because bird-catchers knew they had to take young from the nest soon after hatching if they were to tame them. It was also known by Gessner that crossbills nested incredibly early, often in January or February, whereas many warblers did not breed until May.

The first *systematic* recording of natural history phenomena, however, started with the gardening calendars of the early 1700s. Gilbert White began his *Garden Kalendar* in 1751, reporting both his horticultural activities and the weather. Linnaeus's *Calendarium flora*, which he started in 1757, served as a farmer's almanac, noting the timing of snow melt, swans flying by, underground cellars full of water and swallows drowning themselves – by which he meant disappearing at the end of summer. In 1767 Daines Barrington, who saw the tabulation of

TAB. XLIIII.

Coccothraustes Indica cristata.
The Virginian Nightingale.

Coccothraustes.
The Grosbeak or Haw-finch.

Chloris.
The Greenfinch.

Rubicilla.
The Bullfinch.

Loxia.
The Crosse-bill.

Passer.
A Sparrow.

W. Faithorne sculp.

The crossbill's extraordinarily early breeding season compared with other small birds
was well known to early ornithologists. The crossbill (female, bottom left) often
has eggs in February while there is still snow on the ground. This plate is
from Pepys' copy of The Ornithology of Francis Willughby.

data as synonymous with science, sent Gilbert White a set of printed forms on which to record his observations that in due course became the *Natural History of Selborne*.[28]

An enthusiastic disciple of Gilbert White was the Reverend Leonard Jenyns. Educated at Eton, Jenyns was a meticulous and enthusiastic naturalist. After completing a degree at Cambridge, his father, who was Canon of Ely Cathedral, set him up as curate at Swaffam Bulbeck, adjacent to the family estate at Bottisham, near Cambridge. Like Darwin, Jenyns had a passion for beetles. But as Darwin discovered when they compared collections, Jenyns was from the old school: stuffy and retiring. With plenty of spare time, an interest in natural history and a fastidious manner, Jenyns started a 'calendar of periodic phenomena' in 1820. For twelve years he recorded everything he could relating to the seasonal patterns of plants and animals; the first oak leaf; the first violet and the timing of birds' breeding seasons. Jenyns' scientific strength lay in recognising the fact that biological information like this varied and accordingly, for each phenomenon, he calculated average values from his twelve seasons of observation, but also presented the extremes.[29] For example: the average date on which chaffinches started to lay was 28 April, but across the different years of his study it varied by almost two months between 17 March and 14 May. For the nightingale it varied much less, from 8 to 18 May with an average of 13 May. Jenyns' twelve-year study provided the first quantification of birds' breeding seasons, showing clearly that breeding occurred later in cool springs than in milder ones. Jenyns also recognised that the order in which different species started to breed was similar whatever the spring was like, indicating that each species had its own breeding season and that all species were affected by the weather to the same extent. Jenyns may have been a social fuddy-duddy, but his interest in quantification was well ahead of its time.

A century later the study of birds' breeding seasons became one of David Lack's major research interests. Understanding the evolution of birds' breeding seasons was part of his much broader vision of avian ecology that encompassed their reproductive rates, their lifespan and why bird numbers changed over time. In terms of breeding seasons, the question Lack had to answer was the one Frederick II and John Ray had identified many years earlier: do birds start breeding at the time that results in their young hatching when food is most abundant? It sounds sensible, but it was a tough problem, for measuring the amount of food available in the wild was a daunting task. The German ornithologist Bernard Altum had had a stab at this in the late 1800s, but without much success.[30]

Lack tested the idea by studying great tits, which feed their chicks on woodland caterpillars. By collaborating with entomologists at Oxford who knew how to estimate the numbers of caterpillars in a wood from the amount of *frass* – caterpillar faeces – raining down from the treetops, Lack could reliably estimate the amount of food available to his great tits over the season. The results were remarkable, and convincing: great tits laid their eggs so that their period of chick-feeding coincided exactly with when the caterpillars were most abundant. Subsequent studies confirmed that the timing of breeding in great tits was both heritable and subject to natural selection.[31]

To have young in the nest when food is most abundant requires planning; birds have to start preparing for breeding by nest-building, copulating, accumulating nutrients for egg production and so on, several weeks before food is at its most plentiful. This brings us neatly back to 'stopping' and physiological mechanisms. As Lack recognised, birds like great tits do not start breeding in response to an increase in caterpillar numbers. At the time when they need to start breeding, in March, there are no caterpillars since they are still overwintering eggs at that stage. No: the proximate cue the birds use to know when to start breeding is day length. It is brutally simple and requires no conscious thought. An individual that initiates breeding at a day length that results in it having chicks at the right time will leave many descendants, and the genes for breeding at the 'right' time will be inherited by the next generation. Those individuals responding to the wrong day length – breeding either too early or too late – will leave few, if any, descendants and become mere evolutionary memories.

Aristotle was the first to recognise that pairs of eagles each needed their own space. (From Selby, 1825–41)

6

The Novelty of Field Work

The Discovery of Territory

A little over a century ago you could count the number of bird-watchers on the fingers of one hand. Today there are several million. Considered eccentric, early birdwatchers had a tough time. It wasn't simply that they were regarded as 'odd'; rather, the professional ornithologists were openly aggressive towards them, despising and denigrating their enthusiasm for watching and studying birds in the field. At that time 'real' ornithology comprised taxonomy and classification and was conducted by professionally trained men in museums. With more than a hint of sarcasm, one of them summed up the situation in a few lines:

> Popular ornithology is the more entertaining, with its savor of the wild-wood, green fields, the riverside and seashore, bird songs and the many fascinating things connected with out-of-door-Nature. But systematic ornithology, being a component part of biology – the science of life – is the more instructive and therefore more important.[1]

It was against this tide of ornithological opposition that Edmund Selous decided to swim in the 1890s. Brother of a famous, larger-than-life African big-game hunter, the younger Selous by comparison *did* seem eccentric. Shy, retiring and socially inept, Edmund preferred the company of birds to people. With his book *Bird Watching*, published in 1901, Selous invented both the technique and the name. But what Selous did was rather different from what people consider birdwatching to be today. His passion for watching birds in their natural environment was motivated by science and it was precisely because Selous and his followers encroached on their academic territory that the museum men felt threatened.

Selous's first scientific paper was published in *The Zoologist* in 1899, and described his observations of a pair of nesting nightjars observed at close range from behind an elder bush. This was extraordinary; no one (other than hunters), it seems, had watched wild birds like this before, and the results spoke for themselves. The editor of *The Zoologist* singled out Selous's unique study for praise and *The Saturday Review* declared that Selous had exceeded Gilbert White's observations on the nightjar, referring to him as a 'born field naturalist'.[2]

There was more to Selous than just watching birds. He had a mission, and saw himself as a pioneer. He wanted to change ornithology, to do away with shooting, collecting and classifying birds, summing up his vision thus: 'The zoologist of the future should be a different kind of man altogether: the present one is not worthy of the name. He should go out with glasses [binoculars] and notebook, prepared to see and think.'[3] The idea of giving up the gun in favour of binoculars was ground-breaking indeed. Little wonder the professionals hated him.

Selous was inspired by Darwin's idea of sexual selection, which proposed that differences in the appearance and behaviour between the sexes were due to competition for mates. Males had evolved to be bigger and stronger than females because this gave them an edge when fighting for females. Males were also more elaborately ornamented than females because females considered such males more attractive and preferentially mated with them. Aware that, like natural selection, sexual selection was merely a theory and derived mainly from observations of captive animals, Selous wanted to see whether sexual selection occurred in nature.

After watching birds as diverse as rooks and ruffs, Selous thought it did, and over a period of thirty years produced a succession of books describing his observations and ideas. Reading those books today, however, it is difficult to decide exactly what it was that Selous discovered. Certainly they contain some excellent descriptions of behaviour, and there is little doubt that Selous was an astute observer. But because his prose could so easily become pompous and convoluted it is often difficult to read and harder still to decide whether there are any general conclusions. Here, for example, is Selous on the display of the great crested grebe:

[It] is primarily a physiological reaction to sex stimulus, but that the concentration of the bird's gaze upon the parts thus suddenly exposed, whilst under the influence of such stimulus, leads through inheritance to its prepotent transmission, in which would be embraced such variations, occurring from time to time, as might be included in the sphere of such prepotency – the parts aforesaid.[4]

In the early 1900s Edmund Selous was the first person to observe and record the breeding behaviour of any bird from a hide. His study of the European nightjar, shown here, won him acclaim. (From Meyer, 1835–50)

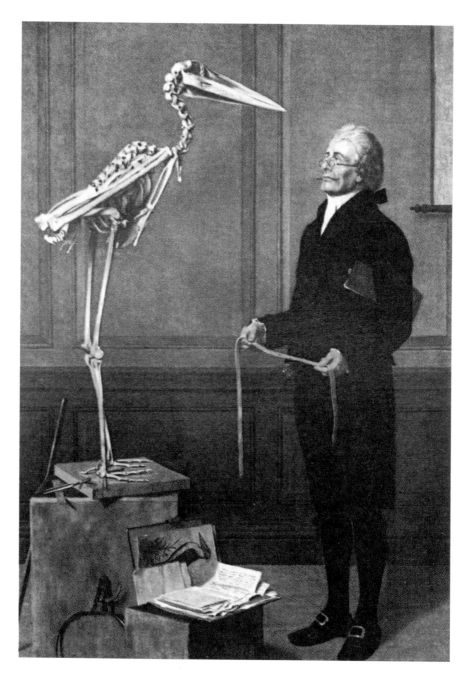

Between 1700 and the early 1900s the only legitimate bird study was museum ornithology, and the museum men mocked and despised field ornithologists like Selous. This painting by Henry Marks from 1873 is entitled Science is measurement.

I wonder whether Selous's writing seemed quite so obtuse to his contemporaries. In an age more tolerant of long-windedness, some probably thought it clever stuff. But one thing is apparent. Selous was inspiring. He was *the* pioneer of ornithological field studies. He started the entire field-ornithology movement. David Lack considered Selous particularly influential in the history of ornithology. But Selous's reputation rests more on what he did than what he discovered. In one sense he discovered nothing other than a method for studying birds in the field; there were no great factual revelations, no sharply focused confirmation of sexual selection and no grand theories of his own – at least none that have stood the test of time.[5]

As for those who further influenced subsequent generations of ornithologists, credit must go to another figure from the early 1900s. Much less well known, Eliot Howard identified and recognised the concept of territory in birds. His *Territory in Bird Life* was one of the most influential bird books of the twentieth century. Now all but forgotten, in its day it caused uproar among ornithologists, revealing territory to be among the most important concepts in bird biology. At one point the world's bird students were said to be in danger of going 'territory-mad', as a result of Howard's ideas. Slightly later the great evolutionary biologist Ernst Mayr expressed his opinion somewhat more soberly, and with good reason, as we'll see, merely commenting that Howard's book had 'caused more discussion among ornithologists than any other recently published'.[6]

A well-to-do amateur bird man, Eliot Howard was born in Kidderminster in the English Midlands in 1873. Educated at Stoke House School, Stoke Poges and Eton, Howard had a passion for natural history throughout childhood. He attended Mason College, which later became the University of Birmingham, to study engineering, afterwards becoming a director of a major steel company in Worcestershire. In his spare time he continued to watch and study birds.[7] Howard was fortunate; his Clareland estate overlooking the River Severn comprised a wonderful range of habitats, from reedy lakes to moorland, with a great diversity of birds, including several species of buntings and warblers. In his thirties Howard decided that the warblers would be the group of birds most suitable for detailed study and he began to make systematic observations of their behaviour in the wild.

They were an excellent choice; as migrants their sexual behaviour started much more abruptly and obviously than it did in the resident species, greatly facilitating his observations:

> Since the spring migration is undertaken for the purpose of procreation, it is probable that the stimulus to that long and arduous journey is the fact that the initial stage of the development of the sexual organs has commenced. Moreover, males arrive before the females, and this fact is of immense assistance in helping us trace out the true meaning of much of their behaviour.[8]

Howard was inspired by Edmund Selous, fifteen years his senior, and, like Selous, was enthused by Darwin and keen to find evidence of sexual selection in the wild. Excited by the mating behaviour of birds, most of Howard's observations concentrate on the early part of the breeding season – pair formation, territory, courtship, culminating in copulation – which he reports with a curious mixture of coyness and unabashed enthusiasm. While he was inspired by Darwin, Howard was disappointed that Darwin should have relied on second-hand knowledge of bird behaviour provided by bird-keepers like John Jenner Weir and Bernard Brent to formulate his idea of sexual selection. Howard was right to be critical: until the early 1900s animal behaviour consisted of little more than anthropomorphic anecdotes of domestic or semi-domesticated animals, epitomised by George Romanes's popular book *Animal Intelligence*. Howard considered such observations worthless compared with the 'impartial investigation of wild nature'.[9]

However, the more Howard watched his warblers the less satisfied he became with what he called Darwin's 'picturesque theory' of sexual selection. He simply could not see males competing for females, nor females choosing between males, as Darwin proposed. Most of the

Blackcaps fighting. Eliot Howard discovered, developed and promoted the concept of territory in the lives of birds. His books contained some of the best illustrations of bird behaviour ever – here two male blackcaps (and an attendant chiffchaff) fight over a territory boundary as depicted by Henrik Grönvold (in Howard, 1909).

fighting he witnessed between males took place before the females had arrived. Once they had, the females seemed singularly uninterested in male displays. In his account of the reed warbler written in 1910 Howard says:

> Those who have paid attention to the habits of birds during the season in which the sexual organs are developing can hardly have failed to notice the battles which are of such frequent occurrence between males of the same species. Darwin believed that the primary object of these struggles was the possession of a female, but ... I am inclined to think that the possession of a territory is of greater importance to the male ...[10]

Compared with Selous's pretentious ramblings Howard's early descriptions of the behaviour of the different warbler species are a joy. Written in straightforward, simple English they convey a fabulous enthusiasm and a superb ability to describe exactly what he saw. Fascinated by what went on in their heads, Howard was especially interested in the differences between individuals – something most other ornithologists glossed over – convinced that understanding the bird's mind would explain their behaviour.

Struggling to come to terms with the bird's mind, Howard sought advice from a professional, Conwy Lloyd Morgan, professor of zoology at University College, Bristol, whose influential book *Habit and Instinct* (1896) contained many observations on birds and included a chapter entitled 'Some Habits and Instincts of the Pairing Season'.[11] Morgan provided steady, sensible advice to Howard, who had received no formal biological training beyond school. Morgan was also interested in the mind, and was the first to ask whether behaviours such as the begging of nestling birds for food or the migratory habits of birds could become instincts – the old nature-versus-nurture chestnut. Morgan's pioneering efforts in animal psychology presented ornithologists, and particularly those like Howard watching wild birds, with a large number of new and exciting questions. For twenty-five years

Morgan mentored Howard, steering him away from dubious anthropomorphism and metaphysics, trying to keep him scientific by reinforcing his skill as an observer. He was only partly successful, and, as David Lack commented long after Howard's death, his obsession with the bird's 'mind' turned him to 'exalted metaphysical regions, where, as yet at least, he has had no followers'.[12] Morgan also recognised that understanding the bird's mind was too complex; too immense; too intractable for Howard (or indeed anyone else) and a century later we are only a little further on in our understanding of it.

The results of Howard's warbler observations appeared in nine separate parts, starting in 1907 with the grasshopper warbler. *British Warblers* is a sumptuous, expensive and superbly illustrated publication. The paintings and photogravures by Danish artist and natural historian Henrik Grönvold were executed under Howard's directions and for me there are few better illustrations of bird courtship and fighting.[13]

The behaviour of wild birds was a long way from mainstream zoology, which in the early 1900s consisted mainly of embryology and comparative physiology. Together with evolution, behaviour was considered 'philosophical natural history', which with its airy-fairy ideas lay well outside what was then thought to be good factual science. But not everyone dismissed it; a few could see field ornithology as an opening horizon. Among these was the printer and publisher Harry Witherby, who in 1907 founded the influential journal *British Birds* that subsequently published much of the debate that erupted over Howard's territory idea.

Another key figure from this era is Julian Huxley, grandson of the zoologist Thomas Henry Huxley, the latter known as Darwin's bulldog for defending natural selection, most notably against Bishop Wilberforce at the famous British Association meeting at Oxford in 1860. A born Darwinian, Julian Huxley took up birdwatching as a hobby as a teenager, and a few years later, inspired by Selous's *Bird Watching*, became interested in bird behaviour and in particular its evolutionary interpretation. After graduating from Oxford with a degree in zoology in 1909, Huxley became a lecturer in zoology there. Still rather uncertain about what his real research focus should be, in the spring of 1912 Huxley spent his Easter vacation studying the courtship behaviour of great crested grebes on Tring Reservoir with his brother. The resultant paper, published in 1914, was a model of clarity and became a milestone in the study of animal behaviour. I remember as an undergraduate being told about this particular study and being inspired by Huxley's clever selection of a bird that did everything out in the open (in contrast to Howard's cryptic warblers), and the fact that in less than two weeks one could make a worthwhile contribution to ornithology.

Like Selous and Howard, Huxley was interested in sexual selection, and his observations of grebe behaviour corroborated something that Howard had also seen in his warblers. Contrary to Darwin's account of sexual selection, which stated that courtship displays had evolved to assist males and females in acquiring a partner, Huxley noted that the displays in his grebes occurred *after* the pair was established, leaving the question of what these displays – with their curious postures, movements and calls – were for. A Darwinian interpretation required that they have a function. Huxley concluded that displays served to cement the bond between partners.[14] It is a logical idea and one that continues to be used today to account for otherwise inexplicable behaviours between members of a pair. But in one sense it is a cop-out. Saying that a particular behaviour helps to maintain the pair bond is merely another way of saying we have no real idea what it is for. The interesting thing is that we now know that the pair bonds of birds are not maintained quite as assiduously as Huxley and Howard once thought. To this day, because no one has thought of a good way of tackling it, the idea of 'the pair bond' remains one of the big unexplored ideas in ornithology. Any ideas?

It was during his early warbler observations that Howard became aware of territory, which he first mentions in his account of the chiffchaff from 1908: 'Breeding territory is a matter of the greatest importance to the males, frequently leading to serious and protracted struggles when two of them are desirous of acquiring the same area.'[15]

As he completed subsequent accounts of the other warblers, Howard gradually refined his ideas, realising that Darwin was wrong in another respect, at least as far as warblers were concerned: males competed for territories not females. Howard discussed his ideas on territory and sexual selection with Morgan, who, recognising this to be a more productive avenue for Howard's energy than his theory of mind, wrote to him on 5 February 1915 encouraging him to write a 'little book embodying your leading conclusions'.[16]

Territory in Bird Life appeared in 1920, proclaiming a very clear message: territory was a general rule, maybe even a *law* of bird life. Howard used Grönvold again, together with another fabulous bird artist, George Lodge, to illustrate his book, both of whom excelled at depicting birds in action. Howard's main conclusions regarding territory were, in brief, as follows. Male birds, through their pugnacity, occupy and defend territories in the spring from other males of the same species, spacing themselves across the countryside and thereby limiting their numbers. A territory ensures a food supply for the young and provides a means by which the bond between partners is maintained. Territory owners are almost invincible in their territory; males compete for territories not females, and male song and displays represent both a threat to other males and an invitation to females.

While *Territory in Bird Life* received some positive reviews, most were rather condescending – the well-known ornithologist the Reverend Francis Jourdain called it an 'attractive and thoughtful little work'.[17] Its immediate effect on the ornithological community was negligible. Howard must have been disappointed, but few great new ideas in science receive instant acclaim. Also, it must have seemed ironic to Morgan that, having encouraged Howard to write the territory book because he felt his ideas had been overlooked, this new book was overlooked as well.[18] But Morgan need not have worried; the effect, when it finally arrived, was greater than even he could have anticipated. It simply took a while. New ideas in biology are often slow to take off, possibly because they require some critical threshold of recognition before researchers decide a topic deserves or

Overleaf: Julian Huxley was a pioneer in the study of animal behaviour: his study of the courtship of the great crested grebe, made in just ten days during Easter 1912, became a classic. (From Selby, 1825–41)

PLATE LXXIII.

GREAT CRESTED GREBE.
1. Adult. 2 Young after 2nd moult.

requires more study. Typically, the younger generation is enthusiastic about new ideas, while the older, more experienced practitioners, having seen new ideas come and go, know there are many blind alleys in research and, like cautious lovers, see how things go before making any commitment. We can gauge the interest in territory by looking at how many papers there were on the topic before and after Howard's book appeared: eleven between 1900 and 1910, fifteen between 1910 and 1920, and, in the decade after its publication, forty-eight, an increase, but a small one compared with what was to come.

In fact it wasn't until E. M. (Max) Nicholson publicised Howard's ideas in his own book *How Birds Live* in 1927 that anyone took any notice.[19] Max Nicholson was one of the great movers and shakers in natural history. Perceptive, impatient, and more of a politician than a scientist, Nicholson was immensely influential in popularising birdwatching and together with James Fisher was particularly effective in bridging the gap between birdwatching and science. Nicholson's book was subtitled A *brief account of bird life in the light of modern observation* and it was the fact that he included an entire chapter on territory that shook the sceptics out of their scepticism. Suddenly territory was something they should be paying attention to: the pulse of interest was quickening.

Nicholson's chapter on territory did more than simply make the professionals aware of territory; it alerted them – drawing them like sharks to blood – to the fact that Howard's theory might be flawed: 'He [Howard] must always have the honour of having been the first to appreciate the importance of territory and to investigate it as fully as it deserves, but whether his personal theory will finally gain acceptance without considerable modification is much to be doubted.'[20] It is difficult to know whether Nicholson was genuinely impressed by Howard's ideas; one wouldn't think so from the way he describes them, but later, when David Lack attacked Howard and aligned Nicholson with him, Nicholson leapt to Howard's defence, saying, 'I am proud to stand in the dock by the side of a man, who, in

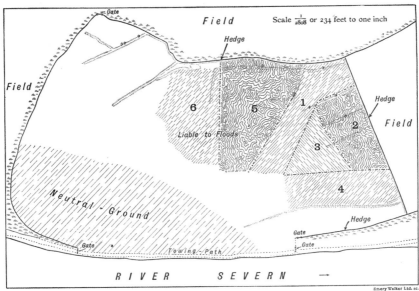

In 1916 Eliot Howard plotted the territory boundaries of six pairs of lapwings on the water meadows of his Shoreland estate, adjacent to the River Severn. (From Howard, 1920; lapwing is from Gould's The Birds of Great Britain, 1873)

my opinion, has done at least as much to further the study of birds as any other living person.'[21]

The professionals started to ask questions. What evidence could Howard muster to support his claims? Aren't there other possible explanations for territory, such as avoiding disease? Are his ideas sound? What exactly does he mean by territory? Is Howard's position as *the* pioneer in territory studies justified? Scientists are obsessed with priority – hadn't others previously suggested some of these same ideas and been overlooked? These were all issues that the professional ornithologists – and it is a star-studded cast – started to ask in the early 1930s.

Among the most vociferous and prolific was Margaret Morse Nice, or Mrs Nice as she was usually known. One of only a few women ornithologists at that time, Mrs Nice was respected for her remarkable life-history studies of the American song sparrow. Informed by her own field observations she was quick to respond to Howard's book and its aftermath, and produced several overviews of the territory concept. I was amazed to learn from one of these that Edmund Selous never read Howard's territory book, yet decided to pronounce on the topic himself, but in such an obtuse way that Nice felt he did more harm than good.[22] I suspect that Selous may have been jealous; his own work had paved the way, but he singularly failed to identify any topic as significant as territory.

The other person who started to notice the idea of territoriality in the early 1930s was David Lack, at that time in his final year of a zoology degree at Cambridge. Together with his father (a surgeon), David Lack published a broad overview of the subject of territory in *British Birds* in 1933. Lack had originally accepted Howard's ideas, but was persuaded by his father, who had become increasingly interested in birds, that the theory was flawed and together they produced a devastating critique of Howard's territory concept, challenging his assertion that territory was ubiquitous and that its main purpose was to ensure an adequate supply of food for the young. David Lack's own view was that territory was 'nothing more than an affair of the male bird, and its real significance seems to be that it provides him with a more or less prominent, isolated headquarters where he can sing or otherwise display'.[23] In other words, territory wasn't a big deal and Howard had overstated his case. But the Lacks were overly harsh; many of Howard's ideas were subsequently shown to be sound. On the other hand, the Lacks' article was immensely important because it gave ornithologists something to get their teeth into, identifying the questions that needed to be answered. As David Lack later said, his 'paper made a stir, and for which I got most of the credit, though my father did almost all the original thinking'.[24] He was being modest; his tutor at Cambridge, Bill Thorpe, described David Lack as 'outstandingly learned and experienced for his age'.[25] A decade later, after some first-hand experience of territory and when Howard's ideas were starting to look more robust than he originally anticipated, Lack was rather more generous, writing in his *Life of the Robin*: 'though partially anticipated in his views by earlier writers, [Howard] must be given full credit for establishing the importance of territory in bird life'.[26]

Before either Margaret Nice or David Lack, the first review of territoriality and Howard's ideas came from the young German biologist Wilhelm Meise, one of Erwin Stresemann's students. This particular paper, published in 1930, passed almost unnoticed in Britain, perhaps because it was in German, but it contained some important information on previous pioneers of territory that properly came to light only when Ernst Mayr took up the issue a few years later. Meise uses Howard's observations on reed buntings to illustrate territory, but also to point out that: 'The results of Howard's research are considered by some authors as something entirely new ... Justice demands that – without intending to diminish the achievements of modern workers – we remind ourselves of Bernard Altum who was ahead of his time by half a century with his views of certain biological problems [including territory].'[27]

Well aware of the work of Altum, Selous, Howard and Nicholson, Mayr had himself identified territoriality as an important question when, in 1929 in New Guinea, he was planning his future research projects. One of the points for research he listed was 'A critical test of the territory theory (Howard, Nicholson)'.[28] When he moved to New York in 1930 Mayr continued his interest in territory, but published nothing on the topic until April 1935, when he produced a paper (in English) identifying Altum's key role.[29] Mayr sent Howard a copy of the paper and in a neat, backwardly sloping hand that betrayed his introspective nature, Howard responded:

> My starting point was, and still is, the congenital foundation which is expressed by a male occupying a region and at the same time becoming intolerant of other males within that region. When a Guillemot occupies a bit of ledge long before an egg is laid and drives away intruders, or a Ruff occupies a station on the meeting ground [lek] and drives away intruders, there is no difference, as far as I can see, in the congenital foundation.[30]

Howard's all-encompassing view of territory was eventually recognised as correct, and is epitomised in what is now considered the best definition of territory, proposed by American herpetologist Gladwyn Noble in 1939 as 'any defended area'.[31]

Reading all these accounts, I was left feeling that the ornithological establishment was irritated and disappointed that it was an amateur who launched the territory revolution. Howard must have been deeply hurt by the Lacks' review. As far as I can tell Howard was totally unaware of the historical literature and had arrived at his conclusions quite independently; the work of Altum, or indeed anyone else who had thought about the topic, was completely unknown to him.

Howard avoided writing further about the history of territoriality, but when Julian Huxley and James Fisher decided to celebrate his career by producing an affordable reprint of *Territory in Bird Life* in 1948, eight years after his death, they felt obliged to add some historical information. Huxley and Fisher were popular, prominent and supremely confident pillars of British ornithology, but probably not the best qualified for writing this particular Introduction (Margaret Nice or David Lack would have been better). With a spectacular career in public life

Huxley was rated one of the cleverest people in Britain in the 1930s.[32] James Fisher was also well known, having done much to popularise the study of birds – his book *Watching Birds* sold over three million copies – and for bridging the gap between professional and amateur ornithologists.[33] He has been described as 'a typical product of the British upper class and like many men of his background, combined an overweening self-confidence, at times bordering on arrogance, with the ability to get on with people of all walks of life'.[34]

In their Introduction to the reprint of Howard's *Territory*, Huxley and Fisher point out that although Aristotle had commented on the seasonal aggressiveness of birds, 'the first indication, in any literature so far discovered, that certain sorts of birds become landowners in the breeding season' comes from Olina's account of the nightingale written in 1622. It was actually the Oxford-based ornithologist and bibliophile W. B. Alexander who discovered this in 1936 on reading Ray's *Ornithology*:

> It is proper to this bird at his first coming (saith Olina) to occupy or seize upon one place as its freehold, into which it will not admit any other nightingale but its mate.[35]

Intriguingly, Alexander also noticed that this paragraph was one of several additions that Ray made to the English edition of the *Ornithology*: it does not appear in the original Latin edition of 1676, suggesting that Ray went back and reread Olina's account.[36] What is more interesting, and rather puzzling, is that, in quoting Olina, Ray omits to say that the nightingale 'ordinarily sings in its freehold', overlooking the crucial link between territory and song.

Giovanni Pietro Olina's sumptuous bird-keeping book *L'Uccelliera*, published in 1622, was produced during the wave of Renaissance enthusiasm for science and objectivity. It was written for (and, indeed, partly by) his employer Cassiano dal Pozzo, who in turn commissioned the book as part of a strategy for getting himself elected to the Accademia dei Lincei, Italy's ultimate scientific society.[37]

Olina's book was far from original, however. Until the nineteenth century, plagiarism was common among all kinds of authors, including those writing about birds, and to create *L'Uccelliera* Cassiano dal Pozzo and Olina borrowed extensively – and often word for word with no acknowledgement – from a previous bird book by Antonio Valli da Todi published in 1601. Valli da Todi in turn had based his work on an even earlier book by Cesare Manzini. We know frustratingly little about either Valli da Todi or Manzini. So who 'discovered' territoriality? To find out I compared the accounts in all three books.

It is clear that the statements about territoriality in nightingales originate from the pen of Valli da Todi rather than from Manzini, who makes no mention of it.[38]

This is what Valli da Todi says about the nightingale:

This furious fight between razorbills over just a few square centimetres of cliff ledge exemplifies the central importance of territory in the lives of birds. Painting by George Lodge (from Howard, 1920).

As soon as it arrives in those parts [of the country], it chooses a freehold, in which place it does not want to have any nightingale entering except its female, and when other nightingales are coming, he starts to sing in the centre of that place, and his nest he makes at a stone's throw away [un tiro di sasso lontano] from where he sings; and he never sings close to the nest, being afraid that serpents or other animals find the nest ...[39]

It was therefore Valli da Todi, rather than Olina, who first reported territoriality in birds, and who recognised the link between territory and song. Significantly, Valli da Todi was a bird-catcher (as well as a bird-keeper) with first-hand experience of watching and trapping wild nightingales. However, I doubt very much whether Valli da Todi was truly the first to notice these things; it seems likely that Greek bird-catchers knew all about territory and song in nightingales as well, but never got round to reporting it to Aristotle.

Aristotle, however, was aware that other birds defend a patch of ground and do so to protect a food supply. In the *History of Animals* he says: 'Each pair of eagles needs a space and on that account allows no other eagle to settle in the neighbourhood. They do not hunt in the immediate vicinity of their nest but go far to find their prey.' Similarly, he says: 'In narrow circumscribed districts where the food would be insufficient for more birds than two, ravens are only found in isolated pairs ...'[40] Writing at about the same time, Zenodotus, the Greek grammarian in charge of the library at Alexandria, said: 'One bush does not shelter two robins', suggesting that the aggressiveness and spacing of robins was also well known from a very early date.[41]

Valli da Todi's account of nightingale territoriality was subsequently repeated, unattributed, by many others. The clue to this rampant plagiarism is in his phrase describing the nightingale's territory as being a radius of a 'stone's throw':

As the nightingales are solitary birds, there is nothing but love and harmony that keeps them together: from the last days of April or the first of May, they start to build their nest. Then the male has the care of choosing a piece of ground and defends this against birds passing by: because it is only a small family that inhabits a small region: another nightingale does not dare to come there without being battered cruelly. Thus, a mated nightingale inhabits a large circular stone's throw.[42]

Modern estimates of the size of a nightingale's territory vary between 0.3 and 0.7 hectares, which is equivalent to a radius of about 31 to 47 m – pretty close to the average stone's throw.[43] Not only did early bird-catchers know about territoriality in nightingales, they also recognised that different species established themselves in different kinds of habitats. Nightingales, for example, were known to establish their territories in cool, shady, wooded areas often near water.[44]

The concept of territoriality in one other bird, the mute swan, was also known at a very early date. In Britain the mute swan is a royal bird, and those on the lower reaches of the River Thames continue to be claimed by the Lord Chamberlain's Office (on behalf of the crown) and two livery companies, the Worshipful Company of Vintners and Worshipful Company of Dyers, who received these rights in the 1400s. Although wild and free-flying, mute swans are carefully protected and during the annual tradition of swan 'upping' are marked for ownership purposes. The birds are relatively tame and easy to observe so it is hardly surprising that their behaviour was familiar to those who protected them. Section 21 of 'The Orders, Lawes and Ancient Customes of Swanns', printed by order of John Witherings in 1632, but based on others at least fifty years older still, reads:

> And yet neither the Master of the Game, nor any Gamster may take away any swanne which is in broode with any other mans, or which is coupled, and hath a walke [i.e. a territory], without the other's consent, for breaking the brood.[45]

Intriguingly, John Ray seems to have been unaware of this account. But even more fascinating is that, having added an extra piece on nightingale territory to the *Ornithology*, he didn't take it any further and say more on the subject. My guess is that not being either a hands-on field ornithologist (no one was at that time) or a bird-catcher, he simply never witnessed territorial behaviour for himself, and didn't recognise its significance. The people that actually got their hands dirty catching birds, the trappers – and swan wardens – were the ones who knew about territory, and there was a gulf between these common people and the educated ornithologist.

In their Introduction to Howard's book, Huxley and Fisher state authoritatively that little new was added to the subject of territoriality during the eighteenth century. But they were spectacularly wrong, for there were several wonderful contributions, two of which are particularly significant.

Marooned for two years on the island of Rodrigues in the Indian Ocean in the 1690s, Huguenot refugee François Leguat watched with obvious delight the breeding antics of a large, flightless, pigeon-like bird known as the solitaire. Leguat noticed that the solitaire – actually a species of dodo – would not tolerate other individuals of its own kind within 200 yards of its nest. The larger male challenged only male intruders, and the female, female intruders. They generated a noisy rattle from a bony knob – the size of a musket ball – on each wing as a keep-out signal throughout incubation and until the young became independent. The solitaire became extinct in the late 1700s. Fisher and Huxley can be forgiven for overlooking Leguat's account, for it wasn't until the 1950s that Edward A. Armstrong rediscovered and publicised it, concluding his own account by saying: 'The name of the bird thus perpetuates the conspicuousness of its territorial behaviour.'[46]

It is more difficult to forgive Fisher and Huxley for overlooking Erwin Stresemann's article published in *The Auk* in 1947 on the extraordinary ornithological observations of Baron von Pernau, Lord of Rosenau, near Coburg in Germany. A wealthy nobleman with a sharp mind and passion for birds, Pernau was well aware in the early 1700s that some species, like the nightingale, robin and redstart, defend a territory and 'do not allow another bird of their kind to come near, except in spring, their female'. In other words, apart from their partner, males kept other males at bay. Pernau also suggests that food is one reason for this:

> The Nightingale is forced, for the sake of her feeding requirements, to chase away her own equals, for if many would stay together, they could not possibly find enough worms and would inevitably starve. Nature therefore has given them the drive to flee from each other as much as possible.[47]

And then on the chaffinch:

> The Chaffinch gives the most pleasure by the males, as soon as the sun in March gets stronger, selecting a special place, just as other birds do, often consisting of a few trees, and by afterwards not allowing another male to show up there. They sing very fervently all day long from the tops of such trees, to induce one of the passing females (that always arrive last on migration) to come down.[48]

Pernau's knowledge of the behaviour of small birds in the wild was extraordinary: he knew that males established territories in March, defended them only against other males, sang to attract females and that unpaired males sang more than paired males. These perceptive observations on territoriality were just part of Pernau's exceptional insight into bird biology, causing one twentieth-century commentator to liken him to a contemporary scientist.[49]

There were several other eighteenth-century accounts of territorial behaviour indicating that by this time it was becoming increasingly well known. In *The Bird Fancier's Recreation* the unknown author says: 'for it is observed by all that know the nature of the nightingale, that he will suffer no competitor'.[50] Eleazar Albin says something similar. The nightingale: 'will not yield to any competitor, either of birds or men'.[51] In a letter to Daines Barrington dated February 1772 (but not published until 1789), Gilbert White refers to the 'jealousy that prevails amongst the male birds that they can hardly bear to be together in the same hedge or field' and 'the rivalry of the males in many kinds

*One hardly expects a shipwreck victim to conduct definitive research, but that
is what François Leguat did in the 1690s on the island of Rodrigues,
with the (now extinct) solitary bird. (From Leguat, 1707)*

prevents them crowding in on one another'.[52] Writing about ruffs in the 1760s Thomas Pennant, who obtained much of his information directly from bird-catchers, stated that 'each male keeps possession of a small piece of ground'.[53]

One of the few eighteenth-century references to territoriality identified by Huxley and Fisher was Oliver Goldsmith's *A History of the Earth and Animated Nature* and Goldsmith may actually have been the first to use the word 'territory':

> The fact is, all these small birds mark out a territory to themselves, which they will permit none of their own species to remain in; they guard their dominions with the most watchful resentment; and we seldom find two male tenents [sic] in the same hedge together.[54]

Huxley might have been the cleverest of men, but he ought to have known that Goldsmith was neither an ornithologist nor a natural historian and his comments about the spring rivalry of birds were unlikely to be original. A brilliant but eccentric hack, ready to translate or write anything – which he apparently did with more confidence than accuracy – Goldsmith is best remembered now as a playwright, poet and novelist. His friend Samuel Johnson, on learning that Goldsmith was to write a book on natural history, said, 'if he can distinguish a cow from a horse, that, I believe, may be the extent of his knowledge'.[55] Not that this stopped him – 'At first glance *Animated Nature* seems like a work of tremendous erudition.' It was one of the most popular natural history books of its time, but it was all stolen from Buffon's *Natural History of Birds* and Brisson's *Ornithologie*.[56] As one biographer noted, Goldsmith had 'the knack of adopting a suggestion and making it ... his own'.[57]

Almost all of Howard's ideas about territory had, as Mayr was keen to point out, been anticipated half a century earlier by Bernard Altum:

Individual pairs must settle at precisely fixed distances from each other. The reason for this necessity is the amount and kind of food they have to gather for themselves and their young ... They need a territory of a definite size, which varies according to the productivity of any given locality ... But many of my readers will ask what is the connection of song ... with the question of territory? ... They sing day after day, in the morning and evening continuously, and by this song the boundaries of the territories are fixed .... Some species of birds, have no definite territories. On a single tower a hundred pairs of Jackdaws may nest ...[58]

Altum also points out that even the expression 'fighting of the males over females' is false:

The males fight to fix the size of the territory, little as they realise the vital necessity of this, and also to select the healthiest individuals for reproduction, but for nothing else.[59]

It is a measure of the poor flow of information that, while Altum's ideas were well known in Germany, they were unheard of in Britain. The similarity in the conclusions of Howard and Altum is remarkable and the only point on which they differ is whether colony-nesting species like the jackdaw defend a territory. Altum, who confines his use of the term to those species that defend a large, all-purpose territory, says they do not. Howard, on the other hand, saw territory as a general law, common to all birds, so that even colony-nesting birds defend a territory, albeit a tiny one (later, Niko Tinbergen and Konrad Lorenz, who both studied colony-nesting birds, confirmed Howard's assertion). But to others it sounded as though Howard wanted to have his cake and eat it, assuming territory to be both ubiquitous and different in different species. The crux of the issue was that Howard failed to provide a definition of territory. Had he done so the controversy would have been resolved, or at least different.

Part of the solution was to consider different types of territory, and Ernst Mayr – clear-thinking as always – was the first to propose a scheme. His classification comprised four types: (a) an all-purpose territory in which all activities take place, (b) an area in which mating and nesting, but not feeding, occur, (c) a mating station, and (d) the area around the nest. Examples of each of these are: (a) European robin, North American robin; (b) Eurasian sparrowhawk; (c) ruff and other lekking species (i.e. those in which males congregate to display and are visited by females for copulation), and (d) most colony-nesting birds, like auks and gulls. After deciding that Gladwyn Noble's definition of a territory as 'any defended area' perfectly encompassed all territory types Margaret Nice subsequently modified and improved Mayr's scheme by adding winter territories (e.g. European robin) and roosting territories.[60]

But it wasn't just definitions that other ornithologists debated. The big-shots were obsessed with three ideas regarding the function or adaptive significance of territory.

First was the role of territory in pair formation. Howard showed conclusively that males did not compete for females, they competed for territories. At first this seemed to undermine Darwin's idea of sexual selection, but in the 1930s Niko Tinbergen resolved this apparent anomaly by pointing out that 'a territory is, to the male, a "potential female"; it is absolutely the same (functionally), if a female is present in the territory at once, or will be present after some time'. Males staked out territories, using song to deter other males, and to attract females, as Altum and Howard recognised.[61]

The second issue was Howard's claim that an all-purpose territory ensured a food supply for the young. Lack, always sceptical, was unconvinced. Because Margaret Nice and Niko Tinbergen had studied different species, they had different opinions regarding territory and food. Others like Lord Tavistock – a keen bird-keeper – were openly hostile to the idea, maintaining that even a human could find enough food for a dozen birds in a tiny portion of a willow warbler's territory, referring to the problem as the 'great food-shortage delusion'.[62] It was obvious that many birds did indeed find a substantial portion of their food during the breeding season within their terri-

Hummingbirds of both sexes defend feeding territories containing nectar-producing flowers, and males defend tiny mating territories. These are the purple-throated mountain gem (top) and white-bellied mountain gem (bottom). (From Salvin and Godman, 1879–1904)

tory; the question was whether food was the *main* reason for defending a patch of ground. The jury is still out. Other ideas proposed by Nice and Tinbergen on the basis of their detailed field observations are that defence of an exclusive area allows pairs to get on with breeding without interference, including sexual interference, from other birds.[63]

The third major concern was the idea that by spacing individuals out across the countryside territoriality prevented overcrowding. The idea was first proposed by the Irish naturalist Charles Moffat in 1903:

> In the course of time, the country – or the parts of it suitable for nidification [breeding] – would come to be completely parcelled out between the birds, each parcel of land belonging to a particular pair; I mean, as against any other pair of the same *kind*. And once this happy state was arrived at, the number of nesting pairs each year would be exactly the same, the number of nests and the average number of young birds reared would be exactly the same; and whether there was a large mortality in winter, or a small mortality in winter, the total number of birds in the country would remain exactly the same.[64]

A clearer statement for the population-regulation role of territory would be hard to find. Moffat thought that there must also be birds without a territory just waiting for a vacancy to arise and supported his contention with reference to the numerous instances of game-keepers shooting one of a pair only to find the remaining bird re-paired within a few days. This view of territoriality maintaining the balance of nature and the best of all possible worlds has great intuitive appeal and took hold in the imagination of many who came later, including Eliot Howard: 'The establishment of territories serves so to regulate the distribution of pairs that the maximum number can be accommodated in the minimum area.' Max Nicholson promoted the same idea in his books, too.[65]

David Lack disagreed completely, adroitly summing up his position in 1943 in his book on the robin:

[T]he idea of 'optimum spacing' involves a fallacy. Natural selection works through individual survival, and this need not result in what is best for the species as a whole. It is the value of the territory to the individual pair and their brood which is the relevant issue.[66]

But the idea was too seductive to go away and later became a major issue in ecology and evolution. Vero Wynne-Edwards, professor of zoology at Aberdeen University, was the main proponent of the optimum-spacing and population-regulation idea. A keen ornithologist, in 1962 he published a huge volume on how animal populations were regulated through social behaviour, including territoriality.[67] His ideas were based on exactly the same mistaken premise that Lack had criticised in Moffat, Howard and Nicholson.

The question of whether territory serves to regulate bird populations was put to rest by David Lack, who pointed out that once you understood that natural selection operates on individuals, not for the good of the species, any effect that territory has in regulating bird populations is a mere *consequence* of territoriality not its function. A subtle but crucial distinction.[68]

During the first three decades of the twentieth century the ornithological pulse began to quicken, but, as far as territory was concerned, by the 1930s it was racing, with more than three hundred papers published on the topic between 1930 and 1940. It was part of a broader change in which museum ornithology and field ornithology were united, a transformation orchestrated initially by Stresemann in Germany and twenty-five years later by David Lack in Britain and Ernst Mayr in the United States. The surge of interest in field ornithology in Britain is largely due to Selous, but it is Howard who made birdwatching *scientifically* respectable.

It is hard to appreciate now how far out on a limb Howard and Selous went and how much they aggravated the British ornithological establishment. That they succeeded in initiating the field study of birds was, I think, because they were amateurs and outside the system. Both men were also loners in their own way: Selous never fulfilled his potential and in old age became cantankerous and unpopular; Howard, on the other hand, was gentle and retiring, with 'a complete absence of intellectual conceit',[69] but also had the common sense and humility to seek scientific advice from someone who was the best at the time, Conwy Lloyd Morgan. Through Morgan's wise guidance Howard identified and developed a major conceptual issue that embraced ecology, behaviour and evolution.

It was, therefore, Howard rather than Selous who helped to launch a revolution; stimulating discussion, raising a multitude of novel biological questions and inspiring subsequent generations of ornithologists to get out into the field. Ernst Mayr finally recognised Howard's contribution, writing to Stresemann in December 1937: 'I am now working toward next year [1938], when I want to have elected Elliot [sic] Howard as Honorary Fellow [of the American Ornithologists' Union].'[70] In 1938 Howard was duly elected, and must have been delighted. In 1959, long after Howard was dead, David Lack on the centenary of the British Ornithologists' Union ranked Howard's *British Warblers* as one of his top five most influential books.[71] Today there are hundreds of field ornithologists, across the world, thanks to Howard.

Whistling bullfinches, trained as youngsters to pipe up to three different folk tunes, were once extremely popular pets. Their affection for their owners and their ability to perfect a tune are unparalleled. (From a painting by K. Schloesser, c. 1890)

7

Choristers of the Groves

Birdsong

In 1943 an eighteen-year-old bird enthusiast by the name of Jürgen Nicolai enlisted in the German army. His period of active service was brief for in a battle against the Russians during Christmas 1944 Nicolai was wounded. After a short spell in hospital he was sent back to the front but en route was captured by the British, who handed him over to the Belgians. As a prisoner of war he endured several years' hard labour working in Belgian coal mines and was released only in 1947. Soon after returning home he spotted an advertisement in a local newspaper offering a hand-reared bullfinch for sale. Purchasing the bird, Nicolai began a life-long relationship with the species and an extraordinary study of the bullfinch's song-learning ability that combined his love of bird-keeping and science.

Fascinated by the almost legendary ability of bullfinches to learn any tune whistled to them by a human, Nicolai began his studies in the 1950s at the University of Mainz, eventually becoming a doctoral student of Konrad Lorenz. Bird-keeping was in his blood: he had bred Harzer canaries as a boy and knew all about the centuries-old tradition of 'whistling bullfinches', in which foresters in the Vogelsberg area of central Germany took the young birds from the nest, hand-reared them and, by whistling a folk tune to them over several months, taught them to sing. The trainers whistled in the same key and at the same pitch to their twenty or thirty avian pupils until eventually the young birds were able to reproduce the tune exactly as they had heard it. The training was laborious and, because the foresters were unaware of the sex of their young birds (half were female and never learnt to sing and not all males became proficient performers), the return on their efforts wasn't always certain. Extremely popular pets, whistling bullfinches were exported all over Europe and, when dealers discovered that birds whistling an English folk tune secured a higher price than those whistling a German or Dutch one, London became a key market. The Victorians loved them. Whistling bullfinches were expensive and owned mainly by the rich and famous. Lizzie Siddal, the Pre-Raphaelite supermodel of the day, had one, as did the wildlife artist Joseph Wolf, who also painted Queen Victoria's bullfinch; and in Russia the family of Tsar Nicholas II owned one.[1]

It wasn't merely their whistling that made bullfinches such endearing pets. The male is among the most gorgeous of all small birds – black head, white rump, steely-blue wings and a beautiful rose-pink breast – their name a bastardisation of the German *Blutfink* or *Blödtfinck* – blood finch. Bullfinches are also extraordinarily affectionate. In nature they appear to be strictly monogamous and pair members remain together throughout the year. If either bird moves out of sight they call to each other and on being reunited the male feeds his mate, regurgitating food from his crop into her open beak. Captive bullfinches also have an incredible ability to recognise individual humans, and presumably other bullfinches, after long periods of separation. When reared by hand, they form a deep and lasting relationship with their owner, equivalent to a pair bond.

The birds that Nicolai used for his experiments were purchased cheaply; they were the foresters' rejects, those that never mastered a tune satisfactorily, but were none the less imprinted on humans. Nicolai soon noticed that, without females, captive males became strongly imprinted on to their owner, and that this fixation took place sometime towards the end of the birds' first year of life. Imprinting was what his supervisor Konrad Lorenz was interested in, so it was hardly

Three views of what is probably a tame male bullfinch, by Albrecht Dürer. Acquired by Philip II of Spain, it is now housed in the Escorial Palace outside Madrid.

surprising that he paid close attention to it. One of Nicolai's fixated males knew exactly when he was leaving the house because it saw him putting on his coat and would start to call. The bird then waited with its crop full of food until Nicolai returned and accepted his 'mate's' courtship feeding, taking the regurgitated food between his finger and thumb.[2]

The bullfinch's remarkable ability to learn one or two separate tunes – very occasionally three – has been known at least since the Middle Ages. 'It is the readiest bird to learn, and imitates a pipe very closely with its voice', wrote William Turner in 1544.[3]

In one sense there was nothing unusual about this. It was well known that other small birds could be trained to sing a novel song. What makes the bullfinch different from all other birds is that, with no song to call its own, it has the cognitive skills and motivation to learn and perform a complex tune taught to it by a human being. To be fair, the bullfinch does have a song, albeit a very primitive one, and about as melodious as a squeaky wheelbarrow, but, unlike other songbirds, this isn't a territorial song for the bullfinch does not defend a territory. The male bullfinch sings only for his partner and does so in an extraordinarily quiet and discreet manner, and it was precisely this gentle and intimate nature of their song, coupled with its purity of tone, that made bullfinches such endearing pets.

All in all, the bullfinch is a very odd bird. Although they are classified as a type of finch, ornithologists are still not quite sure where bullfinches fit in relation to the true finches such as the goldfinch and linnet. During the 1960s, Ian Newton, then at the Edward Grey Institute in Oxford, conducted a detailed study of finches and concluded that in terms of both its behaviour and ecology the bullfinch is dramatically different from all other finch species.[4] First, no other finch shows such an extraordinary compulsion to form an affectionate bond, either with its partner or its owner. Second, as all European bird-ringers know, the bullfinch is extremely highly strung and a trapped bird is likely to drop dead in the hand for no apparent reason. Even in captivity they are delicate. I kept bullfinches for several years and found them to be utterly engaging but at the same time deeply frustrating. Simply moving a bird from one cage to another could send an apparently fit bird into terminal decline. Third, the male bullfinch has extraordinarily small testes, which means it produces tiny numbers of sperm and even these are unusual in their design. These particular reproductive features may be linked with the species' close pair bond, for in a marriage in which the female partner is completely faithful a male needs to produce and inseminate only very few sperm.[5] So far no one has conducted a paternity study of bullfinches, so we do not know whether any extra-pair paternity occurs, but my prediction is that the bullfinch will prove to be perfectly monogamous. Finally, there's the bullfinch's brain. Its ability to learn tunes puts the bullfinch in the top league among song-learners. What is it about its lifestyle that places so much emphasis on this cognitive ability?

Nicolai exploited the tradition of training bullfinches to whistle folk songs to try and answer this question. The overall quest was to discover how birds acquire their songs. During the several years of his Ph.D., completed in 1956, Nicolai used tape recordings and sono-grams (a visual image of their song) to compare the tune whistled by the trainer and that performed by the bird. The bullfinches turned out to be absolute perfectionists, practising endlessly until they could out-perform their trainers. They also had an uncanny sense of how a tune should begin or end.[6] Long after Nicolai finished his bullfinch studies and moved on to other topics, researchers continued to investigate the bullfinch's song-learning talents, revealing among other things an especially intricate coordination between the voice box and the respi-ratory system, enhancing the birds' ability to reproduce whistled tunes so perfectly.[7] There is still much to discover about these unusual birds and another thing I predict is that there will be something extraordi-nary about the bullfinch's brain.

The fact that many different small birds could be trained to sing made it obvious from a very early date that birds acquired their song by learning rather than instinct: nurture not nature. Pernau in the early 1700s said: 'One has to consider that a young bird of any species, which neither hears an adult of its own kind nor has another young around itself, never will attain its natural song completely, but will sing rather poorly.'[8]

Probably independently, Daines Barrington made this same point a century later in his own investigation of birdsong:

> Notes in birds are no more innate than language is in man, and depend entirely upon the master under which they are bred, as far as their organs will enable them to imitate the sounds they have frequent opportunities of hearing.[9]

Most of Barrington's experiments were conducted with linnets, which he used because, unlike bullfinches, the sexes could be easily distinguished (from the amount of white on the wing) from three weeks of age, thus avoiding time wasted on females. Rearing linnets under a variety of species, Barrington found that they invariably adopted the song of their male foster parent. This is an extraordinary phenomenon, as I discovered for myself. I had a pair of siskins that bred in the same aviary as some canaries and when the single male siskin chick matured and first started to sing the following spring, I was startled to hear him pour out pure canary song. Although well aware of this kind of effect, I could never quite get used to the sight and sound of a siskin singing the song of another species. The early pioneers of song-learning in birds, like Pernau and Barrington, must have been equally fascinated by this effect. Nicolai recounted a particularly striking instance in which one of his male bullfinches adopted its canary foster father's song, and when this bullfinch was paired to a female bullfinch their offspring also sang only canary song, as did their offspring in turn![10]

The linnet chicks that Barrington used in his studies were all three weeks old when he started to raise them, so he could not exclude the possibility that they had already heard their father. Because of this Barrington knew that his experiments were not watertight. Ideally, he said, the chicks should be taken much earlier, and, best of all, before hatching, so they have no opportunity to hear their own species' song.

*Two common European cage birds of previous times: the goldfinch (bottom),
kept for its melodic and varied song, and the chaffinch (top), often employed
in singing contests. (From Frisch, 1743–63)*

But he also acknowledged that rearing a bird from hatching was extremely difficult and 'the odds against it being reared are almost infinite'. However, he came across two cases where birds had been reared from two or three days old that confirmed his views. The first was a linnet owned by a London apothecary; the bird's *only* vocalisation was the phrase 'pretty boy'; the second was a goldfinch Barrington heard as he was walking past a house on the Welsh borders, which he first thought to be a wren. The goldfinch had been reared from a young age and – he assumed – had heard a wren and adopted its song. Barrington was amused that the owners were completely unaware that the goldfinch wasn't singing its natural song.[11]

Barrington was also well aware that birds had to practise their song: his main intellectual rival, Comte de Buffon, had described how fledgling nightingales start to warble as soon as they begin to feed themselves, and that their song developed gradually, until by December it was at full power. Subsequent research on chaffinches – one of a handful of model species for song studies – revealed that young birds go through distinct phases of the learning, starting with 'subsong', the quiet, babbling, featureless song that matures into 'plastic song', which is more like the typical song but lacks the organisation of the final 'full song'. Subsong sounds as though the bird is practising and does not want to be heard, and that is exactly what is happening. The bird is literally singing to itself – listening and refining its own performance. Listening to themselves is crucial since birds that have been deafened as adults continue to sing a normal song, while those deafened during the subsong phase never develop a proper song.[12]

To impress on his readers the importance of learning in song acquisition Barrington drew an analogy with the appreciation of music by humans:

> I am also convinced (though it may seem paradoxical), that the inhabitants of London distinguish more accurately, and know more on this head, than of all the other parts of the island taken together.

His reasoning was that Londoners heard better quality music via the opera than their country relatives and that this was – by degrees, he says – communicated to the fiddler and ballad-singer in the streets. He is careful to say that he does not mean to imply that there is any inherent difference in the musical organs of the inhabitants of the country and of Londoners but only: 'that they have not the same opportunities of learning from others who play in tune themselves'.[13]

Clearly Barrington believed in the absolute role of nurture, learning or environment – whichever you chose to call it – in the acquisition of song by birds. Once he had adopted a particular point of view he was a hard man to persuade otherwise, as we have already seen with his attachment to the idea of hibernating swallows. Most of those who followed seem to have accepted without question the idea that songbirds learnt their song. Buffon, however, had an altogether more open mind and as a result was probably more perceptive. In the introduction to his *Natural History of Birds* Buffon states categorically: 'Sweetness of voice and melody of song are qualities, which in birds are partly natural, partly acquired.' He acknowledges that birds can learn certain tunes, but doesn't elaborate on the 'natural' part of birdsong, although I suspect he means that the core characteristics of song are innate.[14]

The innate nature of the core features of birdsong was finally demonstrated in the 1960s when Cambridge zoologist Bill Thorpe reared chaffinches in auditory isolation; although their song bore little resemblance to a normal chaffinch song, it was none the less recognisable as belonging to a chaffinch because of its tonal quality. In fact, evidence that certain aspects of birdsong are innate and have a genetic basis had been available since the 1800s when German canary-breeders focused their efforts on producing a bird that sang a distinctive rolling song: the roller canary. Through a combination of selective breeding and early training, the breeders in the Harz Mountains produced a canary with a unique soft rolling song.[15] Although there are also some wild birds, such as the European great tit, whose song is innate,[16] the species studied by Barrington – the chaffinch, goldfinch, skylark and

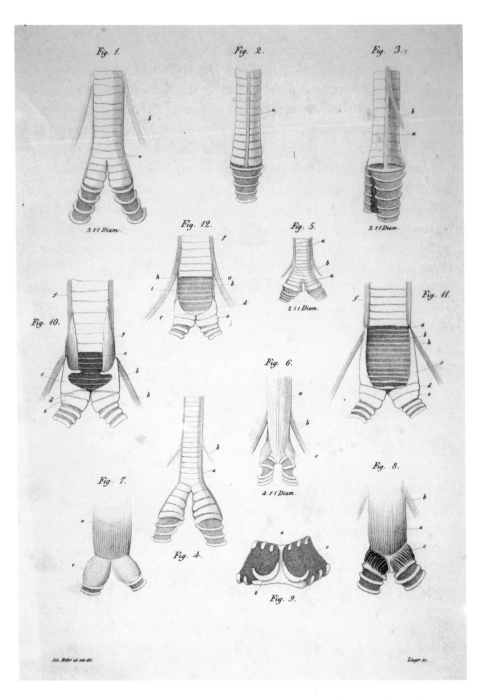

The birds' voice box or syrinx is complex in its structure and extremely variable between species. The examples here are from Müller et al. (1878).

nightingale – were all ones that learnt their song and accounts for why he was so adamant about his results.

Barrington was also intrigued by how birds actually made their songs, and in the 1770s asked the famous surgeon Sir John Hunter to examine the voice box of a range of bird species for him:

> I procured a cock Nightingale, a cock and hen Black-Bird, a cock and hen Rook, a cock linnet, as also a cock and hen Chaffinch, which that very eminent anatomist, Mr Hunter, F.R.S. was so obliging as to dissect for me, and begged that he would particularly attend to the organs in the different birds, which might be supposed to contribute to singing.[17]

Hunter reported that in general the 'muscles of the larynx' were more developed in species with better voices, like the nightingale, and larger in males than in females. He also described the air sacs[*] in birds and speculated about their role in song:

> How far this construction of the respiratory organs may assist birds in singing, is deserving of notice; as the vast continuance of song, between the breathings, in a canary-bird, would appear to arise from it.[18]

Around 1600 Aldrovandi reported how a bird whose head had been removed was still able to vocalise, clearly locating the source of sound in the bird's body rather than its head. This simple if somewhat unpleasant observation – which must have been well known to poultryfarmers – made a mockery of the view held by Aristotle and most other early writers that the bird's tongue is (as in humans) the main organ in sound production. At the same time, however, Aristotle also acknowledged the importance of the voice box in human speech, although the parallels between the human voice box (larynx) and avian voice box (technically the syrinx) were not recognised until much later.[19]

---

[*] The air sacs in a bird are an extension of its respiratory system. They form a system of thin-walled, transparent sacs that act as bellows to bring air into the bird, but are not involved in gas exchange.

Charles Darwin's ingenious grandfather, Erasmus, constructed a sensational speaking machine that mimicked – albeit crudely – the workings of the human voice box and the mouth. A pair of bellows forced air across a silk membrane, which, together with a pair of leather lips, resulted in the machine speaking a few simple words. Erasmus Darwin's machine provided important clues as to how the vocal apparatus of birds and humans might work, but it was obvious that the complex and beautiful songs of birds required something much more elaborate.[20]

The bird's voice box is both immensely complex and highly variable in design between different species. The first inklings of how it might work were discovered only in the 1950s when researchers had the bright idea of passing air through the syrinx of anaesthetised birds. These studies indicated, as in Erasmus Darwin's speaking machine, that sound was produced by the air passing over the membranes inside the syrinx. Later, in some astonishing research, scientists placed tiny cameras inside the windpipes of live, singing birds and showed that it was actually the connective tissue inside the syrinx rather than the membranes themselves that created the sound.[21]

A human being without a tongue cannot speak since the tongue is vital for shaping and perfecting the sounds produced in the larynx. Early ornithologists assumed the same to be true for birds and, notwithstanding his observation that a headless bird could still vocalise, Aldrovandi and others still thought the tongue played a crucial role in sound production. Indeed, on looking into the mouth of a caged nightingale he was perplexed by his failure to see a tongue: 'which circumstance excited in me considerable wonder, that the little bird should have such sweetness of song and such vibration of voice, and yet be without a tongue, unless it were perhaps concealed in its throat'. Aldrovandi was correct in thinking the tongue concealed – it was simply lying flat at the bottom of the mouth – for all birds possess a tongue but in most it plays no part in their vocalisations.[22]

In a perverse way the obsession with the tongue as a vocal organ led to the unnecessary and cruel practice of mutilating those birds – like

TABVLA LXV.

I.I.R.pinx.R.

I.G.F.sc.R.

The starling's astonishing ability to mimic both other birds and the human voice
made it a popular cage bird. It was erroneously assumed that if birds like this
were to speak, it was necessary to 'loosen' their tongues. This picture from
Schaeffer (1779) includes an image of the starling's tongue.

the starling and the magpie – that were trained to talk. As a boy in the 1950s I caught a young starling hoping to teach it to speak and recall being told firmly by my uncle, a farmer, that I should split its tongue if I wanted to be successful. One account recommended:

> [cutting the] black part off with sharp scissors, not farther than the co-loured part, then the tongue is already made round. Then both white dots on both sides of the tongue should be cut off and given more room by cutting a small part of the membrane which links the tongue to the throat. After that the bleeding parts are treated with a little unsalted butter, and this operation is repeated three times, i.e. each week once, for instance on Friday. After the string of the tongue has been clipped loose the talking lessons could start.[23]

The earliest reference I found to tongue mutilation in birds was from 1601 in Valli da Todi's bird-keeping book, but I suspect the practice may be much older. Not least because the idea of slitting, trimming or loosening a bird's tongue almost certainly came from the ancient practice of dealing with tongue-tied children. In cases where the membrane under a child's tongue extends too far forward, restricting the tongue's movement and impeding speech, the solution was to cut the membrane to 'loosen the tongue'. For both children and birds the operation sounds deeply unpleasant. I tried to see how widespread it was in birds, and in a dozen books published between 1728 and 1889 in which there was mention of tongue-clipping, all authors without exception agreed that it was cruel and completely unnecessary: 'To slit their tongues, as many people advise and practise, that the bird … may talk the plainer, is of no service; they will talk as well without, as I have found by experience; as will likewise magpies, and other talking birds'. Sadly, the same was true of children too.[24]

It has since become clear that in those ultimate mimics of the hu-man voice, the parrots, their large, fleshy tongue really does play a vital role in vocalisation. Making use of some monk parakeets killed as part of a control programme in the southern USA, where they are considered a pest, a team of biologists performed an ingenious experiment in which they replaced the syrinx with a tiny, hearing-aid speaker and played sounds through the dead birds' vocal tracts and recorded the sounds emitted through the beak. To their astonishment they found that by moving the tongue 'a fraction of a millimetre makes a larger difference than an A and an O in human speech' and may help to explain the extraordinary ability of parrots to mimic human speech.[25]

The production of song or other avian vocalisations depends, of course, on more than the syrinx or, in the parrot's case, on tongues. As John Hunter recognised, the entire respiratory system is also involved, as are (albeit less directly) the testes (i.e. testicles) and the brain. The link between the testicles and song was well known in humans. The habit of castrating pre-pubertal boys to retain the purity of their high-pitched voice started with the development of European opera in the sixteenth century, although for centuries before that human castrati had been used as harem guards. The effect of castration could be startling and the voice of a castrato was likened to that of a nightingale. Opera audiences loved them and shouted '*Viva il coltello!*' – Long live the knife! From about 1500 poor parents offered up their sons for castration – barbarously performed by barber-surgeons – in the hope that they might become famous singers. Noting that the castration of cockerels caused them to cease crowing, Daines Barrington asked:

> why this operation should not improve the notes of a nestling, as much as it is supposed to contribute to the greater perfection of the human voice. To this I answer, that castration by no means insures any such consequence; for the voices of much the greater part of Italian eunuchs are so indifferent, that they have no means of procuring a livelihood but by copying music, and this is one of the reasons why so few compositions are published in Italy, as it would starve this 'refuse of society'.[26]

Obviously uncomfortable with the existence of eunuchs, Barrington speculated that in birds castration would result in a failure of the

Overleaf: An unexpected vocalist: despite its familiar harsh chatter the magpie has a soft, bubbling, rarely heard song and can also be taught to speak. One that I knew could recite entire nursery rhymes. This painting is by Johann Walther c. 1650.

syrinx muscles to develop. He attempted to test the idea by persuading 'an operator' to castrate a six-week-old blackbird, but the unfortunate bird died. Barrington could 'only conjecture with regard to what
might have been the consequences of it', although as it turns out he
was correct: a bird's testes and its syrinx are linked by testosterone.

First identified in the 1930s, the role of testosterone in song production is crucial. With the testes as the main source of testosterone,
castration inevitably halts or reduces singing, as Barrington would
have discovered had his blackbird survived. Castration in young human males eliminates testosterone and prevents puberty – hence the
falsetto voice. In birds, the crucial test of this idea was performed by
Fernando Nottebohm in the 1960s on a young male chaffinch whose
song was still incomplete. After being castrated the bird stopped singing, but when it was given testosterone a couple of years later (and
long after chaffinches have normally finalised their song) the bird began to sing and copied a tape of chaffinch song played to him. The
conclusion was that 'the critical period for learning was not just determined by age, but by the stage of the bird's own neural development,
or directly by testosterone'.[27]

One of Nottebohm's students, Art Arnold, went on to explore
these questions in more detail, using the zebra finch (rather than
the chaffinch which was unavailable in North America) as a study
species. His experiments produced totally unexpected results, for, in
contrast to the chaffinch, zebra finches without testes sang almost
completely normal songs! Critics suspected that Arnold had botched
the castration operation, leaving fragments of testes to regrow, but as
Arnold told me: 'I did a fairly exhaustive job of looking for testicular
remnants ... and was confident that the castration was successful.'[28]
The real reason castrated zebra finches continue to sing is almost unbelievable. As well as producing testosterone in their testes like the
chaffinch and other birds, the zebra finch also produces testosterone
in its brain! This extraordinary phenomenon may be an adaptation to
the zebra finch's nomadic, arid-country lifestyle where they have to be
ready to respond opportunistically and quickly to breeding chances. In

contrast, in other songbirds testosterone is necessary only for the final learning phase during which song is perfected.[29]

When testosterone was first produced synthetically in the 1930s unscrupulous bird-dealers injected female canaries with it to make them sing so they could pass them off as males. The way that testosterone caused females to sing was discovered by Nottebohm in the 1970s. To his and everyone else's amazement, he found that in those areas of the female's brain concerned with song, testosterone caused certain nerve cells to double in length. This was a monumental discovery – not just for ornithology, but for biology as a whole and neurobiology in particular.

Through their ability to learn complex vocalisations by means of auditory feedback, songbirds provide a wonderful model for human speech learning. Findings from the studies of songbirds have had a dramatic impact on studies of brain development and function in humans and much of what was first discovered in birds turns out to be true in mammals as well. Briefly, in the brains of songbirds the vocal centres are typically larger in males than females and larger in species with larger song repertoires – Barrington would have been delighted by this. Each spring as the male comes into breeding condition, under the influence of sex hormones both his testes and the song regions of his brain undergo a huge increase in size. The changes in the brain, which include the creation of *new* neurones, were once thought to be impossible since it had always been assumed that cells in the brain neither replace themselves nor regenerate – hence the inability of the brain to repair itself after injury. The realisation that the old dogma was wrong led to a revolution in brain research, revealing that in humans and other mammals certain areas of the brain *are* capable of producing new neurones. Together with new developments in stem-cell biology these findings promise new treatments for a range of neurological disorders such as Parkinson's and Alzheimer's.[30]

'Lord, what music hast thou provided for the saints in heaven, when thou givest bad men such music on earth!' So wrote Izaak Walton in the 1600s of the nightingale's song.[31] Birdsong is among the most aesthetically pleasing and uplifting of natural experiences, causing great speculation regarding its purpose.

The title of a recent scientific book on birdsong, *Nature's Music*, encapsulates and celebrates the way we perceive the vocal outpourings of birds. One of that book's editors, Peter Marler, is currently the most senior figure in a dynasty of birdsong researchers that was established in the 1950s. The book's focus is science, but it provides a stimulating summary of the many areas of current research into bird vocalisations. Some earlier books on birdsong or songbirds have similarly evocative titles, including *The Choristers of the Groves* and *The Sweet Songsters of Great Britain*.[32] In pre-Darwinian times the powerful, emotive effect of birdsong was the ultimate expression of God's creative powers. Intriguingly, Edward Armstrong in the 1960s suggests that apart from the occasional 'dainty expressions of enjoyment' by Pliny, Irish scribes and Chinese poets, the aesthetic appreciation of birdsong started only after the Renaissance. If it is true (and I doubt it) Pierre Belon must have been among the first, for throughout most of his enormous encyclopaedia Belon writes with cool scholarly detachment, but in describing the song of the nightingale he becomes almost poetic, asking: 'Could there be a man lacking judgement who would not have any admiration hearing such melody coming out of the throat of such a wild little bird? ... Has it ever had a master, who has taught it the science of such perfect music?'[33]

Darwin wasn't insensitive to the beauty of birdsong either, but his goal was different: to understand the advantages it conferred and how it had evolved. One of the books that helped shape his thoughts was William Gardiner's *The Music of Nature* from 1832, which explored – in a typically long-winded Victorian manner – the links between birdsong and music.[34] Darwin eventually wrote:

> Music arouses in us various emotions, but not the more terrible ones of horror, fear, rage &c. It awakens the gentler feelings of tenderness and love, which readily pass into devotion ... It likewise stirs up in us the sense of triumph and the glorious ardour for war ... It is probable

that nearly the same emotions but much weaker and far less complex, are felt by birds when the male pours forth his full volume of song, in rivalry with other males, to captivate the female.[35]

And birdsong has indeed inspired music. Mozart kept a pet starling precisely for this purpose. When he bought the bird in 1784 it was already an accomplished performer and, remarkably, could whistle part of Mozart's own Concerto in G major (K. 453), which may, of course, be why he bought it in the first place. Who had trained the bird was a mystery for that particular piece had not then been performed in public. The bird lived with Mozart for three years and when it died in 1787 he honoured it with an elaborate funeral. Mozart's biographers dismissed the starling affair as an eccentricity that had little influence on the composer's creativity, but, as the birdsong researcher Meredith West points out, none of these biographers was an ornithologist and none of them had lived, as she had done, with a pet starling. West started to look at the music Mozart composed around the time he owned the starling, and one piece, entitled 'A Musical Joke' (K. 522), stood out from the rest. Musicians consider it a 'marvellous and malicious prank' and a 'parody of poor composition'. But West heard in it 'the autograph of a starling ... the fractured phrasing of the entire serenade, tiresome repetitions and eccentric ending, sounding as though the instruments had ceased to work'.[36]

Meredith West studied starlings by accident. She was investigating how song developed in a completely different species, the brown-headed cowbird, and acquired a starling as a companion for a cowbird, but of the two the starling turned out to be the more interesting. The ability of starlings to mimic and incorporate other sounds, including the songs of other birds, human speech and mechanical noises, into their songs is well known. Pliny relates how the young Caesars taught starlings and nightingales to speak, and writing in the late seventeenth century Nicholas Cox says that the starling 'is kept by all ranks of people, and taught to pipe, whistle or talk'.[37] It was also obviously a favourite of Johann Bechstein, nineteenth-century author of one of the most popular bird-keeping books of all time, in which he says:

The starling becomes exceedingly tame in confinement ... It learns, without having its tongue loosened to repeat words, whistle airs (a power shared by the females also), and to imitate the voices of men and animals and the song of birds. It is, however, very uncertain in this respect, as it not only soon forgets what it has learnt, but mixes up old and new lessons together ...[38]

This mixing-up of lessons that Bechstein felt was a shortcoming of the starling is what West witnessed in her pet bird, and, far from being a defect, is exactly what starlings do: editing and rearranging their library of sounds in an endless and dynamic repertoire. West's bird incorporated human voices into its songs, but in a typical mixed-up starling manner, so that the phrase 'we'll see you later' became 'see ya later', 'see you' and 'we'll see'. The bird was also highly selective in the phrases and sounds it incorporated into its song. Cocking its head to one side and evidently listening, the bird ignored many common phrases and specialised in rare sounds. It subsequently uttered sounds as though to gauge their effect and West concluded that starling mimicry is a way of finding new stimulation. The endless acquisition of new sounds is simply part of the birds' biology, and in fact it all made sense when later research revealed that male starlings with a larger song repertoire enjoy greater reproductive success.[39]

Many people have kept starlings as pets, including the behavioural ecologist and ornithologist John (later Lord) Krebs, who for several years had a starling in his office in Oxford, where it learnt to give a perfect imitation of him answering the telephone. In an attempt to train the bird to sing Krebs left a tape of Tamino's flute piece from *The Magic Flute* playing over a weekend. When he came back on the following Monday morning the bird gave a perfect rendition of the first few bars.[40]

Mozart used musical notation to describe the sounds his starling made, as did earlier musicians such as Athanasius Kircher,[41] but they all knew that this was an inadequate way to render birdsong. And in fact the main obstacle to studying birdsong was the difficulty of

The yellowhammer is one of a handful of birds whose song can be rendered phonetically, as A little bit of bread and no cheese. This striking portrait is from Marcus zum Lamm and dates from the mid-1550s. (From Kinzelbach and Hölzinger, 2001)

In 1650 Athanasius Kircher produced an encyclopaedia of music that included the songs and calls of birds in musical notation. Here we see his attempt at the nightingale's song, as well as the simpler calls of the fowl, quail, cuckoo and macaw.

seeing' and measuring it. When Barrington made his assessment of the quality of song produced by different species in the 1770s, his only technical equipment was his ear and his brain. His was the first attempt to quantify birdsong, scoring species according to different criteria, including mellowness, sprightly notes and compass of their song. None the less, this was still an extremely subjective exercise and, not surprisingly, few agreed with his rank order of singers.

Describing birdsong by the use of adjectives such as sonorous, sharp or gay, or likening syllables to particular sounds, as Bechstein does at great length for the chaffinch,[42] is hopeless and conveys little or nothing of the true sound. Somewhat better are phonetic renderings, such as *A little bit of bread and no cheese* for the yellowhammer or *Take two then Taffy* for the woodpigeon; and in North America, *Drink your teeee* for the towhee or *Pleased, pleased, pleased to meetcha* for the chestnut-sided warbler. The main advantage of such mnemonic rhymes is that they are easy to remember and you don't need to be able to read music. On the other hand there are very few birds for which such rhymes exist and they too are subjective, as is neatly demonstrated by rendering of the cockerel's crow. In England this is the familiar *cock-a-doodle-do* (Shakespeare's version was *cock-a-diddle-dow*), *corcorico* in France, *kikeriki* in Germany and *kokke-kokko* in Japan.[43]

A few phonetic renderings are more accurate, including the chaffinch's song (which occurs in four phrases):

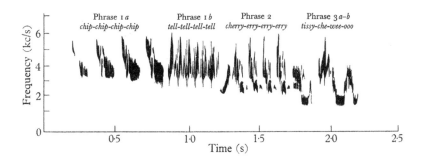

The invention in the 1940s of the sonograph, which provides a picture of sound (with duration along the horizontal axis and pitch or frequency on the vertical axis), revolutionised the study of bird song. Here the match between the sonogram and the phonetics of the chaffinch's song is clear. (From Thorpe, 1961)

How much more difficult to render the complex song of the nightingale, yet this is exactly what the poet and bird-lover John Clare was able to do. Clare spent a lot of time listening to nightingales and he catches better than anyone the quality and cadence of the birds' song:

> *Chew chew chee chew chew*
> *Chew – cheer cheer cheer*
> *Chew chew chew chee*
> *Tweet tweet tweet jug jug jug.*[44]

And so on.

The subjective assessment of birdsong ended with the invention of the sound spectrograph (also known as the sonograph) in the 1940s, an instrument that produces a picture of the frequency and duration of sound. Recognising the tremendous potential of this new technology, Bill Thorpe acquired one in 1950, and in doing so transformed the study of birdsong.

Thorpe trained as an entomologist and started out studying the role of imprinting and learning in the choice of food plants by insects. Over time he became more interested in learning than in insects and, as Robert Hinde noted, by the 1940s: 'Thorpe felt that the key problem was the relation between instinctive and learned behaviour, and that birds, with their stereotyped movement patterns and marked learning abilities, would provide the ideal material.'[45]

A keen birdwatcher, it wasn't difficult for Thorpe to switch from bugs to birds, and in 1950 he established a field station at Madingley just outside Cambridge. Setting this up involved some territorial negotiations with David Lack, who had recently become director of the Edward Grey Institute in Oxford, for Thorpe was concerned that if he invested in aviaries in Cambridge Lack would not do the same in Oxford. In fact, because Lack's group ended up focusing on wild birds there was no problem. Thorpe appointed Robert Hinde, who told me: 'Bill Thorpe and I had to learn everything [about keeping birds] from scratch. We were lucky in having a wonderful lab assistant,

The chestnut-sided warbler, a North American species whose song can be rendered
Pleased, pleased, pleased to meetcha. *(From Audubon's* Birds of America, *1827–38)*

Gordon Dunnett, who was or became a fancier himself ... I spent a fair amount of time visiting fanciers in Cambridge – I remember being surprised at how many bathrooms had been turned into bird rooms. We bought some old parrot aviaries from the Duke of Bedford ... In the early days we kept a wide variety of species – especially finches.'[46]

Thorpe, inspired by the recent results of a Danish study, decided the chaffinch would be the perfect study species.[47] Instinct and learning were the focus of the new field of animal behaviour and, after reviewing what was known about learning in birds, Thorpe was able to turn a lot of old ideas on their head and identify the questions he needed to address. Peter Marler joined the group in 1951 and promptly conducted a wonderfully detailed study of chaffinches both in the field and in captivity. This monumental study was similar in many ways to one that Robert Hinde had completed earlier in Oxford on the great tit under Lack's supervision. Written at a time when knowing your study species inside out was considered essential before starting experiments, these accounts are still worth reading today. Hinde and Marler both went on to illustrious careers; Hinde broadened his interests to include primate and human behaviour, and Marler stayed with birdsong.[48]

While Bill Thorpe revealed in unprecedented detail *how* birds acquire their songs, what his research did not address was *why* birds sing. There had been many ideas but no consensus, elegantly summarised by Darwin who said: 'Naturalists are much divided with respect to the object of the singing of birds.'[49]

They were still undecided in Thorpe's day. Some considered birdsong an outpouring of self-expression – emotional release – with no specific function. Others believed it to be God's gift to mankind. To others birdsong was practice – but didn't specify for what, thereby dodging the issue of function. Some thought that male birds sang in order to entertain their partner during the tedious incubation process: 'It is only the female who incubates, such as only the female of the canary does: while their males keep on singing in order to please them in the nuisances of their nests.'[50] Buffon was under no illusions about the object of birdsong:

The natural tones of birds ... express the various modifications of passion ... The females are much more silent than the males ... In the male it [the song] springs from sweet emotion ... And what proves that love is among birds the real source of their music is, that after the breeding season is over, it either ceases entirely, or loses its sweetness.[51]

For Darwin birdsong was a product of sexual selection. As Buffon noted, song was one of the main differences between males and females and was clearly important in reproduction. Like other sexually selected traits, such as elaborate plumage or the antlers of stags, song has no obvious survival value. Just the opposite, in fact. Since song is energetically demanding and renders males more conspicuous to predators it is more likely to reduce than enhance their survival. As Darwin saw it, the only way traits like this could evolve was if the advantages they provided through mating offset any survival disadvantages. He referred to the idea as sexual selection and thought of it in terms of some individuals leaving more descendants than others (analogous to natural selection in which some – better adapted – individuals survive better than others). Darwin envisaged sexual selection operating in two ways: through competition between males, and through female choice. The idea of males competing for females was not new (his grandfather had commented on it), but the concept of female choice as part of sexual selection was Darwin's brainchild.

It is just possible that Darwin got the idea of female choice from Johann Bechstein's *Natural History of Cage Birds*. First published in 1795 in German, Darwin owned one of the numerous English editions in which Bechstein discusses the male canary's beautiful and powerful song – the *raison d'être* for owning canaries, he says. But he also states, almost in passing, that female canaries prefer the best singer. In suggesting that the male canary's song was about mate attraction, Bechstein was simply repeating – albeit in a more focused manner – what many bird-keepers had suspected for centuries.[52]

The other idea for the function of birdsong – male competition – was also well known to bird-catchers, who exploited it in their trade:

Indeed it has often been experimented, that, when one nightingale in a cage is placed near another on the land, and close to the place where the latter has his nest, but so that they could not fight with each other, one or the other sings himself to death, because they search for very special notes and are driven too high, in order to surpass each other, and to display their ardour to each other. In general one knows well that both the large as well as the little birds do not bear each other's vicinity, when they sing, in particular not near their nest …[53]

The more precise link between the frequency of song and the stage of the breeding cycle of an individual nightingale pair provided an even better clue to what was going on:

Almost coinstantaneously with the hatching of the Nightingale's brood, the song of the sire is hushed … No greater contrast can be imagined, and no instance can be cited which more completely points out the purpose which song fulfils in the economy of the bird, for if the Nightingale's nest at this early time be destroyed or its contents removed, the cock speedily recovers his voice, and his favourite haunts again resound to his bewitching strains.[54]

However, the resumption of song by male nightingales if the breeding attempt failed helped neither to identify the function of song nor distinguish between the two mains ideas of love and aggression.

George Montagu, famous for his *Ornithological Dictionary* and infamous for abandoning his wife for the female artist who illustrated it, appears to have been among the first to have recognised the dual purpose of male song: an expression of love and male rivalry. He also appreciated the way the amount of song distinguished mated from unmated males for both human listeners and females of the same species, advertising a male's availability. However, Montagu appears not to have seen Bechstein's interpretation of canary song and did not consider the possibility that the *quality* of a male's song might form the basis for female choice. None the less, Montagu's pronouncement on the dual function of birdsong marked the beginning of the modern era of birdsong studies with respect to its function. Despite this, almost a century later the great Alfred Newton in his own *Dictionary of Birds* wrote that 'No comprehensive account of the song of birds seems ever to have been written'. Newton – a hard-core museum man – mentions a few key papers, including Barrington's, and although he acknowledged the importance of song in what he disparagingly called the 'economy of birds', other than re-emphasising the link between song and the early stages of reproduction – what he calls love or lust – doesn't venture any unambiguous function for song.[55]

The definitive tests of Montagu's dual-function hypothesis for birdsong were a long time coming. In the early 1900s Wallace Craig conducted a detailed study of the vocalisations made by doves. Not song, strictly speaking, but the pigeon equivalent, and Craig's now forgotten paper was full of original insights, including the fact that the male dove's cooing stimulated the female's reproductive system.[56] Then, in the 1970s, Don Kroodsma at Rockefeller University, New York – with no explicit reference to either Craig or Bechstein – set about testing the idea that females preferred and were most stimulated by the best singers. Kroodsma played isolated female canaries either elaborate or simple songs, and recorded how quickly they built nests, as a measure of how stimulated they were by the songs. The females that heard the elaborate songs built their nests more rapidly than those experiencing the poorer songs, confirming that females could discriminate between song types but also responded most strongly to complex songs.[57]

Twenty years later researcher Eric-Marie Vallett and his colleagues at the University of Paris were able to establish exactly what it was in certain songs that female canaries found so stimulating. Embedded in some songs and barely perceptible to the human ear were some rapid two-note trills – uttered up to seventeen times a second – that females found extremely stimulating. Played in isolation these sexy syllables – as Vallett called them – had a particularly potent effect on the female's brain and caused her to solicit for copulation. Sexy syllables comprise a high- and-low frequency sound produced by the two sides of the syrinx and because such complex sounds are difficult for males to produce they may provide a reliable cue to the quality of a male. In other words, by choosing a male on the basis of his ability to create these sexy sounds, a female acquires a high quality partner. Studies of other bird species also show that males with larger song repertoires are preferred by females.[58]

The territorial function of birdsong was not tested explicitly until the 1980s, when John Krebs conducted a conceptually simple but logistically complex experiment with great tits. In a small area of woodland near Oxford, Krebs plotted the territory boundaries of all great-tit pairs. He then recorded their songs before capturing all the males and replacing them by small loudspeakers positioned around their old territory boundaries. In some cases the speakers played the original owner's song, in others they played a 'control' sound – a penny whistle – and in the third group there was no sound at all. Krebs predicted that if song was a keep-out signal, the silent territories should be reoccupied by new territory-seeking males first, followed by the penny-whistle territories and lastly by those territories broadcasting great-tit song. His predictions were beautifully borne out: song clearly did serve as a keep-out signal to other great tits. Krebs went further and repeated the experiment with songs that differed in repertoire size, to test the prediction that larger repertoires were more effective keep-out signals than small repertoires. One of my doctoral contemporaries, Ruth Ashcroft, worked as a research assistant for Krebs on this project and when I cynically quizzed her about the results she admitted that she too had been sceptical, but that the results were quite clear: larger repertoires were more effective. Krebs attributed this to what he called the Beau Geste effect: a larger repertoire created the illusion to would-be recruits that there were many more great tits than there really were (readers of Percival Wren's now unfashionable novel *Beau Geste* may appreciate the analogy). Other researchers subsequently found that great tits with larger repertoires are older and are better breeders – possibly because they have better territories – but are also preferred by females.[59]

George Montagu – reviled for his unconventional personal life – is vindicated and would have been thrilled by these results: a better song ensures both a better territory *and* a better partner for females. Song, as we will see in the next chapter, is a prelude to sex.

The first known portrait of a zebra finch (this is the Timor race rather than the more familiar Australian form), from Vieillot (1748–1831). A model bird species for researchers, the zebra finch will be the first passerine bird whose genome is sequenced.

8

A Delicate Balance

Sex

V ery, very occasionally bird-watchers come across a bird with a puzzling appearance, with male plumage down one side of its body and female plumage on the other. Looking like the victim of a cruel trick, such birds are referred to as 'half-siders'. These extraordinary individuals are better known among bird-keepers, although they are still extremely rare. Out of 15,000 zebra finches reared over twenty-five years by one research laboratory in the United States only a single half-sider occurred. This particular bird became a star and featured in the journal *Science* as the first half-sider ever to be studied using modern molecular tools. With male plumage on its right side and female plumage on its left, the bird behaved as a male: it sang a normal song, courted and copulated with females that laid eggs that failed to hatch. Internally, on its left side the bird possessed an ovary and on the right side a testis that produced sperm. Examination of its DNA revealed the bird's brain to be genetically male on the right side and genetically female on the left.[1]

A half-sider is a hermaphrodite, technically known as a lateral gynandromorph,* and is usually genetically male on the right-hand side of the body and female on the other, although occasionally the reverse

---

* From lateral meaning side; gynandro meaning female-male; and morph meaning appearance.

is true. Half-siders are not confined to birds; they sometimes also occur among mammals and Aldrovandi included an illustration of a very
rare human gynandromorph in his book of monsters of 1642. In the
past, particularly, half-siders – human or otherwise – always attracted
attention, especially from the Church for which any form of sexual
deviancy was considered a crime.[2]

In 1474 a cockerel accused of laying eggs was burnt at the stake in
front of an immense crowd in Basel. On slicing open the cock prior to
putting it to the torch, the executioner found three more eggs, confirming that the bird was indeed an inauspicious creature whose fate
was fully justified.[3] Being burnt at the stake was a fairly typical medieval response to something that was both biologically inexplicable
and religiously subversive. Of course, at that time honest biological
interpretation was irrelevant; what mattered was the bird's symbolic
significance. No one in medieval times attempted to read nature objectively; instead they preferred to read into it 'an elaborate, arbitrary
and artificial significance'.[4]

Hens that changed sex were also bad news, portents of disaster:

> A whistling woman and a crowing hen
> Are neither good for God nor men.

Worse still, an egg laid by a cockerel was thought to hatch a basilisk
or cockatrice – part bird, part serpent – whose glance would kill. Mentioned in the Bible, basilisks continued to invoke morbid fear through
the Middle Ages and they feature in both Gessner's and Aldrovandi's
encyclopaedias. The weasel was the only creature capable of safely
combating a cockatrice – secure in the knowledge that it could protect
itself by eating rue leaves – and there is a lovely medieval misericord
in Worcester Cathedral depicting a battle between two weasels (each
carrying a rue leaf in its mouth) attacking a cockatrice.[5]

While we might marvel at the medieval ability to interpret anything out of the ordinary as a bad omen, belief in egg-laying cockerels
persisted among some naturalists well into the eighteenth century.

Hermaphrodites: two 'half-sider' Eurasian bullfinches. On the left is a bird that is male on the left-hand side of its body and female on the right; the other bird shows the opposite pattern. Half-siders have one testis and one ovary. These hermaphrodites appear to be more common in the bullfinch than in any other species. (From Kumerloeve, 1987)

Gallus monstrificus barbatus cornutus
ocreatus cauda anguina in cuius
extremitate est flocus prope Vropigium
autem ubi adheret corpori habet quoddam
tuberosum rotundum colore albido
crista palearibusq plumosis.

Half reptile, half cockerel, a cockatrice was thought to be the product of
a cockerel's egg. This one is from Aldrovandi (1600–1603).

One of the earliest scientific attempts to investigate an egg-laying cockerel took place in Copenhagen in the 1650s. On royal command, the physician and anatomy professor Thomas Bartholin dissected the offending bird in the presence of King Frederick III, but found nothing unusual; the bird appeared to be a male with normal testes and its vas deferens swollen with semen, but no ovary or oviduct. The egg apparently laid by this cockerel was slightly smaller than an ordinary hen's egg, but normal in every other respect and, as Bartholin acknowledged, there was no way of knowing whether it had been produced by the dissected cockerel, or by some other bird in the farmyard where it had been found. The investigation was inconclusive, but Bartholin continued to be intrigued and in 1670 acquired another 'cock's' egg that he artificially incubated along with some normal eggs. Again the results were ambiguous; none of the eggs hatched, those of the hens' proved to be sterile and the 'cock's' egg contained albumen, but no yolk.[6]

Thirty-seven years later another dubious cockerel came under scientific scrutiny, this time from the French savant and royal surgeon M. Lapeyronie. The appearance of small, yolkless eggs in a particular farmyard was attributed to a cockerel which M. Lapeyronie thought might be a hermaphrodite, but on dissection was found to be a perfectly normal male. The unusual eggs continued to appear, however, and were finally traced to a hen that also happened to crow like 'a hoarse cock (*un coq enroué*), only more violently'. The hen was duly dissected and found to have its oviduct compressed by a bladder of 'serous fluid' as large as a man's fist and it was assumed that the unfortunate hen crowed because of the pain caused by passing the egg. On opening the eggs laid by this sick female, Lapeyronie found they contained no yolk, only albumen, some of which was twisted into the form of a tiny serpent – which is precisely what the farmer who owned the bird assumed it to be. On the basis of his findings Lapeyronie proudly dismissed the notion of cock's eggs and basilisks, but also provided clear evidence that an egg-laying hen could – possibly through ill health – sometimes assume male characteristics, such as the ability to crow.[7]

In fact it had been known since Greek and Roman times that gender among chickens was a highly labile trait. Aristotle noted that elderly cockerels occasionally developed a sort of egg: 'substances resembling the egg ... have been found in the cock when cut open, underneath the midriff where the hen hath her eggs, and these are entirely yellow in appearance and of the same size as ordinary eggs'.[8] It was also well known that removal of the gonads would greatly alter the appearance and behaviour of either a cock or a hen: 'The fact is that animals, if they be subjected to a modification in minute organs [i.e. the gonads], are liable to immense modifications in their general configuration.'[9]

The castration of farm animals is an ancient and widespread tradition. The practice varied in its execution but was uniformly brutal. According to Aldrovandi:

> Among the Ancients, roosters were castrated in a way quite different from ours. They used to burn with a hot iron the hind part of the bowels or loins or spurs: Roosters are castrated in the lower part of the bowels, which falls down when they have coition. If you burn this with two or three hot irons, you make a capon.[10]

Pliny also mentions that cockerels were castrated either by burning the genitals with a hot iron, or – unbelievably – burning the birds' feet. Aldrovandi describes the process employed in his day, around 1600:

> Our farm wives pull out the testicles through the posterior parts after making a small incision [near the anus] with a knife. The wound is large enough to admit a finger above the genitals under the septum where the testicles adhere and is sufficient to draw them out one by one. When the testicles are removed they sew up the wound with thread and scatter ashes over it. They also cut off the rooster's crest in order to remove all his virility.[11]

This was a dangerous operation, and, as Aldrovandi says: 'roosters die if some mistake is made in the process of castration ... Every care must be taken to remove each testicle, for if one is left in the rooster it crows, sings, seeks coition, and grows less fat.' And the purpose of this vital butchery? A great meal: 'A platter full of this bird aids the stomach, soothes the breast, makes the voice sonorous, and fattens the body.' As though this weren't enough, capon meat was also considered to have aphrodisiac properties.[12]

Many early biologists commented on the dramatic effects of castration. William Harvey, for example, noted that after being deprived of their testicles cockerels lost all vigour and fecundity, and Aldrovandi described how capons sometimes assumed the hen's role of rearing chicks. He also asks rhetorically why capons are 'so remarkably afflicted with gout and roosters are not'? and later answers his own question: 'Because the capon has little heat [lust] and much gluttony.'[13]

It was usually males that sacrificed their gonads, although female castration also occurred occasionally and with the same objective of making a better eating bird. One consequence of losing their single ovary* was that hens typically developed cocky tail feathers. Researchers later showed that if female castrates – referred to as *poulards* – then had a testicle surgically implanted in their abdomen, or were given a shot of testosterone, they developed a cockerel's comb as well.[14] Even more remarkably, in some instances where females have their left ovary removed either by surgery or by disease, the right ovary, which is normally vestigial, roars into life, not as an ovary as we might expect, but as a fully functioning sperm-producing testicle![15]

The fact that a female bird deprived of its reproductive organs should adopt male plumage at its next moult, but castrated males did not revert to female plumage (even though their *behaviour* was feminised), puzzled biologists for a long time. Indeed, the whole business of how an individual becomes male or female is a long-standing problem in biology, and because birds clearly have a remarkable

---

* Birds typically have only a single functional ovary (the left one) – probably due to lack of space and as a weight-saving adaptation for flight (see also chapter 2); the right ovary is vestigial.

potential to change sex they played a crucial role in resolving this mystery.

In 1780 the surgeon Sir John Hunter related to the Royal Society the curious case of a peafowl that changed from a female to a male: 'Lady Tynte had a favourite pyed pea-hen, which having produced chickens several times; having moulted when she was about eleven years old, she astonished the lady and her family by the feathers peculiar to the other sex, and appearing like a pyed pea-cock.'[16] Fascinated by reproduction, Hunter made several detailed studies of hermaphrodites, and was among the first to distinguish between primary and secondary sexual characteristics. The former refer to the genitalia, the latter to structures like the peacock's tail or a stag's antlers that externally distinguish males from females, which Darwin was later to incorporate into his theory of sexual selection.

The ornithologist William Yarrell, a friend of Darwin's, was the first to reveal that disease was the most likely cause of females transforming themselves into males. Of seven masculinised hen pheasants Yarrell dissected, all had diseased ovaries. Yarrell also tested his idea more directly by experimentally removing the ovary of a hen pheasant and found that, once the bird had recovered from its operation, it began to crow and grew a comb and spurs, like a male, confirming that the ovary was essential if a female was to remain female.[17]

There were other important experiments performed around this time, including some by Arnold Berthold, curator at the Göttingen zoo during the mid-1800s, who discovered that testicular extracts could restore a castrated cockerel's sexual attributes, and others by Claude Bernard who recognised that the gonads produce certain 'internal secretions' (i.e. hormones) that are carried in the blood to target organs elsewhere in the body.[18] Despite these vital discoveries, there was still confusion over hermaphrodites and birds that changed sex, until the early 1900s when Francis Crew decided to make the topic of intersexuality the focus of his life's work.

As director of the first institute of animal-breeding in Edinburgh, Crew had unparalleled access to animal-breeders, whom he asked to

*In the late 1700s Lady Tynte had a peahen that – much to her surprise –
changed sex, despite having reared several broods of young over the preceding years.
(From Nozeman, 1770–1829)*

send him their sexually confusing animals. One particularly striking case in 1923 involved a Buff Orpington hen that had laid eggs for three years, and then changed its appearance to become a rooster that went on to sire two chicks. On dissection, the bird was found to possess two functional testes and a shrivelled ovary – possibly destroyed by a tumour – just as Yarrell had found in his pheasants.[19]

Science is often sexist and, perhaps not surprisingly, until recently it was widely assumed that male hormones controlled male sexual traits – like a cockerel's plumes. The truth is that many male characteristics are determined not by the male hormone testosterone, but by the *absence* of the female hormone, oestrogen.[20] Typically, males do not secrete oestrogen (or at least not very much), and consequently produce male plumage. Females secrete oestrogen and develop female plumage. A female with a diseased ovary that ceases to produce oestrogen reverts to the default situation of male plumage, which is what happened to Lady Tynte's gender-bending peafowl.

In a similar way, the most likely explanation for the medieval egg-laying 'cockerels' was that these birds were actually female – that is, genetically female – but with some kind of ovarian disease suppressing the production of oestrogen so that they assumed the outward appearance of a male, with elongated tail feathers and crowing behaviour, but retained the ability to produce eggs.

The control of male breeding plumage by means of the hormone oestrogen is typical of birds like chickens, ducks, pheasants and the peafowl in which males moult into showy plumage only for the breeding season and assume a drab plumage for the rest of the year. In contrast, with birds like the house sparrow, where males retain their distinctive appearance throughout the year, the differences in plumage of the two sexes is determined genetically and is completely independent of hormonal secretions. This is easily demonstrated, for if house sparrows of either sex are deprived of their gonads their plumage remains unchanged.

There is, however, one interesting exception to the oestrogen rule: a wading bird whose name simultaneously identifies it and describes the male's spectacular breeding plumage – the ruff. In contrast to all other birds described so far, the elaborate breeding plumes and curious facial warts of the male ruff depend critically upon the presence of testosterone.

In the nineteenth century Italian bird-catchers performed a barbaric operation on captive male ruffs. By removing all the feathers from one side of the body and allowing them to regrow, they could generate a bird that looked completely different on one side from the other: dull winter plumage on the unplucked side and a bright, highly patterned summer plumage on the other. A different kind of half-sider.

The ruff is not a species one generally thinks of as a cage bird but ruffs have been kept in captivity since the Middle Ages, not for their aesthetic qualities, which are considerable – at least among breeding males – but in order to fatten them for the table. Technically the name ruff refers to the male; the female is the 'reeve'. The sexes are distinguished by the male's greater size, and during the breeding season by his ruff of neck feathers, head tufts and facial warts. This extreme difference in size and appearance of males and females is typical of birds with a polygamous mating system and is the product of intense sexual selection. Males compete furiously for females; some are spectacularly successful and fertilise a number of different females, while other males leave no descendants at all.

Because ruffs are such good eating they have been trapped for centuries and bird-catchers provided much of our early knowledge of ruff biology. Trappers were well aware of how variable the males were in their plumage, some having ruffs of black feathers while others were chestnut or white. One of the pioneers of ruff behaviour was Edmund Selous, who, having read a popular account of their mating dance by the Dutch schoolteacher Jacob Thijsse, travelled to the island of Texel to see them for himself in the spring of 1906. As he had hoped, Selous obtained abundant evidence that females chose which males to copulate with, just as Darwin's theory of sexual selection predicted: 'The selection on the part of the reeves is most evident. They take the initiative throughout and are the true masters of the situation.'[21] It is

The extraordinary variation in the elaborate plumes and facial warts of male ruffs during the breeding season advertises their status on the mating grounds. These images are from J. L. Frisch (1743–63) (top); J. A. Naumann (1795–1803, vol. 2) (bottom left); and Thomas Pennant (1768) (bottom right).

a tribute to Selous's observational skills that he also noticed that different-coloured males seemed to adopt different mating tactics. Fifty years later two Danish students confirmed Selous's suspicion by showing that some males (mainly dark-ruffed individuals) defended tiny territories, just 30–60 cm in diameter on the lek, while others, mainly those with white ruffs, appeared content to share with a territory owner.[22] It was a Dutch ruff researcher, Lidy Hogan-Warburg, who made the most detailed observations of this behaviour, calling the dark-plumaged males 'residents' and the white ones 'satellites', recognising that the latter were sexual parasites on the residents. In other words, the white-ruffed birds hung around in the territory of a dark-ruffed male in the hope of securing the occasional copulation from females attracted to the resident.[23]

In the early 1800s George Montagu acquired some ruffs from a Mr Towns, a Lincolnshire ruff-catcher and fattener, and kept them in his menagerie. Montagu tells the story:

> We took the trouble of carrying several of these birds from Lincolnshire into Devonshire, in hopes of keeping them for several years, in spite of the opinion of Mr Towns, that they could not be kept alive through the winter. These beautiful little partners in our carriage, were taken out of their basket twice a day, and put into a corner of the room wherever we stopped for refreshment, and with a few chairs and a piece of canvas hung over them, reaching the ground, they were perfectly contented, and appeared as happy as fighting and eating could make them: and in such a situation they passed each night of the journey. The last of these birds lived in confinement four years, and several for two or three years, which gave us an opportunity to observe more minutely their manners and change of plumage: and we noticed that their annual changes never varied; every spring produced the same coloured ruff and other feathers; but the tubercles on the face never appeared in confinement.[24]

The fact that males developed the same colour plumes year after year indicated to later ornithologists that the ruff's plumage colour was under genetic control, even though the mechanism was unclear. Then, in the 1930s, Dutch ornithologists interested in the role of hormones discovered that male ruffs castrated before the breeding season failed to develop either ruff or wattles and remained in winter plumage, indicating that testosterone was essential if the males were to develop their breeding plumage.[25] A final piece in the ruff jigsaw was discovered relatively recently when Canadian researcher David Lank established a breeding colony of ruffs in captivity. Lank's pedigrees indicated that the male's breeding plumage might be determined by a single gene on one of the regular chromosomes (i.e. on an autosome), that is, not on a sex chromosome. The female ruffs, of course, have no fancy breeding plumage, but if Lank's idea was right about the plumage gene being on an autosome then males and females should possess the *same* genes even though their effect is apparent only in the males. To visualise these otherwise invisible plumage genes Lank performed an ingenious experiment and administered doses of testosterone to his captive *female* ruffs.

The effect was dramatic. Within forty-eight hours the females started to display typical male behaviours; within a week they had established a lek, their body weight increased as it does in male ruffs (to avoid having to feed while focusing on sex); they began to defend typical territories and five weeks after the testosterone injection the females had started to grow a ruff of elongated neck feathers. Lank's hypothesis about the mode of inheritance predicted that the females' new plumage would be of the same type as their brothers' (with whom they were likely to share the same genes) and this is exactly what he found, confirming both the role of testosterone in plumage development and the genetic basis for the plumage differences and behavioural strategies between male ruffs.[26]

There's a final twist in the ruff's hormonal tale. For the vast majority of ruffs, the difference in size between males and females is so great that even when they are in their dull, identical winter plumage, their sex is obvious. Occasionally, however, bird-catchers on the Dutch polders catch migrating ruffs in winter plumage of intermediate size whose sex is therefore unclear. These birds are a puzzle for, although Theunis Piersma and Joop Jukema recently showed from DNA samples that they were genetically male, in captivity they never develop the typical breeding plumage of male ruffs. It turns out – remarkably – that these birds represent a third genetic male morph and one that appears to be a female mimic in terms of its plumage, but with very large testes. By posing as a female, this third morph makes its living on the display grounds of other males, sneakily mating with any females the territory owner attracts.[27] The physiological processes that allow this third morph to look like a female but still retain functional testes remain to be established, but seem likely to involve uncoupling the link between testosterone and male plumage development.

Virgin birth is another sexual aberration the medieval Church would have found ambiguously fascinating. In many animals – including certain birds – females can reproduce without sex, with no contribution from the male whatsoever, a phenomenon referred to as parthenogenesis.* The Swiss naturalist Charles Bonnet discovered parthenogenesis in the mid-eighteenth century during his studies of aphids.[28] He was one of several researchers to realise that even in species that normally needed sperm, eggs could often be induced to start development on their own: silkworm eggs plunged into hot water for a few minutes will do so, as will the eggs of frogs and toads if pricked by a pin. Such observations merely added to the mystery of reproduction and seriously questioned whether males had any role at all.

In the 1930s poultry researchers at the Agricultural Research Center, Beltsville, Maryland, began creating a strain of turkeys they hoped would be all things to all men: rapidly maturing, highly productive and very fertile. By the 1950s the wonder bird – known as the Beltsville Small White – was established and during a routine

* Partheno = virgin; genesis = birth.

fertility check in 1952 one of the researchers, Marlow Olsen, noticed something unusual. Although long since isolated from any males, the eggs of some of these female turkeys appeared to be fertile. It was not so much that some eggs had started to develop – parthenogenesis in chickens had been discovered in the nineteenth century by the embryologist J. Oellacher and, although extremely rare, was well known – it was the high incidence of apparently fertile eggs among his turkeys that Olsen found surprising. No less than 16 per cent of almost 1,000 eggs examined by Olsen showed some kind of embryo development. But Olsen also knew that in the few birds in which parthenogenesis occurred – chickens, turkeys and pigeons – development was *always* abortive: the cells started to divide, but did so in a disorganised way and *never* resulted in a normal embryo that survived to hatching.

Olsen was intrigued none the less, particularly since some of his female turkeys seemed to produce more parthenogenetic eggs than others, suggesting that the phenomenon might be genetic in origin. A major study was launched, partly to understand what was going on, but also in the hope that there might be some commercial application for asexual turkeys. Sure enough, the tendency to produce automatically developing eggs was heritable, and by carefully selecting the most promising birds Olsen increased the proportion of parthenogenetic eggs from 16 to 45 per cent in just five generations. Not only that, this artificial selection resulted in an increase in the numbers of parthenogenetic embryos reaching an advanced stage of development. Finally, in 1955, Marlow Olsen and his colleagues became the proud parents of the first parthenogenetic turkey chick. In due course there were several more, but these virgin-birth turkeys tended to be rather weak and had to be helped from the shell; many died soon after hatching, but as the researchers became more adept at caring for them, several of their special babies survived and some even reached sexual maturity. During the entire twenty-year project (which, it turned out, had no commercial application) a total of 1,100 parthenogenetic turkeys were hatched. Since they were produced with no contribution from

LXVII.

WILD TURKEY *Meleagris gallopavo* : Male : Female : Young

A bird that can reproduce without sex: the turkey is one of just a handful of birds in which parthenogenesis (virgin birth) is known to occur. This illustration by Alexander Wilson (from Wilson and Bonaparte, 1832) shows a family of wild North American turkeys, with the male on the left.

the male, they were expected to have only a single set of chromosomes (that is, from their mother), but in fact these birds all had a duplicate set, indicating that somehow a doubling of chromosomes occurred at a very early stage of their development. The parthenogenetic turkeys were always male, as their testes confirmed, and they had no interest in sex. Some, however, produced small amounts of sperm which, when artificially inseminated into normal turkey hens, produced normal heterosexual offspring.[29]

Wild turkeys have been around for millions of years and Olsen speculated that a few parthenogenetic birds may have been hatched naturally, but because they were so sickly compared with normal birds they would rarely have survived. Ornithologists have usually dismissed parthenogenesis as a poultry–pigeon artefact, a quirk of domestication with no relevance to other birds. I felt much the same until I recently discovered parthenogenetic eggs in captive zebra finches, suggesting that this reproductive curiosity may occur among all birds occasionally. It also made me wonder whether, with twenty years of effort, I might create a parthenogenetic zebra finch.[30]

John Ray would have struggled to explain hermaphrodites and virgin birth among birds, and as far as I can tell was quite unaware of such things. He must have known about the ability of birds to change sex, however, from reading Aldrovandi and from his correspondence with Thomas Browne, both of whom wrote about it, but Ray chose not to discuss it. Instead he focused on a more obvious example of God's providence – the creation of equal numbers of males and females:

> One thing necessary to the conservation of the species of animals; that is, the keeping up constantly in the world a due numerical proportion between the sexes of male and female, doth necessarily infer a superintending Providence. For did this depend only upon mechanism, it cannot well be conceived, but that in some ages or other there should happen to be all males or all females, and so the species fail. Nay, it can-

not well be thought otherwise, but that there is in this a providence, superior to that of the plastic or spermatic nature, which hath not so much of knowledge and discretion allowed to it, as whereby to be able to govern this affair.[31]

That most species produce roughly equal numbers of male and female offspring had been known since the earliest times. The Greeks had many ideas for what determines the sex of a child, all of which involved what we now call environmental sex determination, including such things as whether the embryo had implanted in the left or right horn of the uterus, or whether the sperm had come from the right or left testicle. Ray thought it improbable that the equal production of male and female offspring depended solely upon what he called 'mechanism', because if it went wrong and only one sex was produced, the species, he said, would be doomed. In a world made perfect by God and with no notion of either evolution or extinction, this wasn't unreasonable, but we now know that sex is indeed determined by 'mechanism'. Our understanding of evolution also allows us to imagine what immense selection must have operated to produce a mechanism of sex determination that operates so efficiently.

The genetic basis of sex was not established until about 1900 when Karl Correns, one of the rediscoverers of Gregor Mendel's work, showed in a plant with separate sexes that half the pollen was male-determining and the other half was female-determining. Researchers subsequently discovered exactly the same system in mammals; half the sperm result in male offspring, half in females. The secret lay in the sex chromosomes.

In mammals – as in ourselves – males have two sex chromosomes referred to as X and Y, and when males make sperm either one of these sex chromosomes is included, so sperm are either X-bearing or Y-bearing. The female mammal has two identical sex chromosomes, both X, so when ova are made they are all X. Since the male mammal's two sex chromosomes are different, he is referred to as the

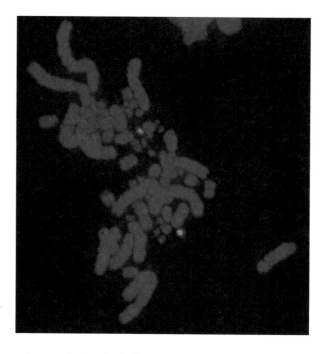

The full complement of zebra finch chromosomes. The two pink spots indicate a gene (on two different chromosomes – one from each parent) involved in disease resistance. (From D. Griffin and B. Skinner)

heterogametic sex,[*] and because the female mammal has two X chromosomes, she is the homogametic sex.[†] Sex is determined by which kind of sperm fuses with an ovum; a Y-bearing sperm fusing with an ovum (X) gives rise to a male (XY) offspring, while an X-bearing sperm fusing with an ovum (X) will give rise to a female offspring (XX).

The situation in birds is different from mammals in that the female is the heterogametic sex. The sex chromosomes in birds are referred to as Z and W, and females are ZW and produce Z- or W-bearing ova; males are homogametic (ZZ) and all their sperm are Z-bearing. In birds, therefore, it is the female that determines the sex of her offspring.

The existence of sex chromosomes in birds was suspected in the early 1900s and one of the first people to pursue this was the cranky

* Hetero = different; gametic = of the gametes.
† Homo = the same.

megalomaniac William Bateson. While Bateson tackled the problem in chickens, his long-suffering research assistant Florence Durham studied canaries. Either Miss Durham or the canaries were more tractable than the chickens and in 1908 she produced conclusive evidence that the female was the sex-determining sex. However, it was another few years before anyone could actually see the sex chromosomes. This was because, in contrast to mammals, birds possess a very large number of chromosomes, many of which are very small, so sorting out which were the sex chromosomes was tricky. By the 1930s it was known that the male fowl was ZZ, but it remained unclear for some time whether the female was ZW or merely Z and lacked a W chromosome.[32]

Of course, the fact that half the gametes that one sex produces are either male or female neatly accounts for why, overall, half the offspring are male or female. But why a system with an *equal* sex ratio evolved in the first place is one of the major unanswered questions in biology. Also why there are only two sexes, rather than three or more, is unclear. Because sex is determined at random when the female bird's sex cells divide to form ova, it is entirely due to chance whether a female produces a male or a female egg, and occasionally, purely by chance that females produce a run of offspring all of the same sex. One captive Eclectus parrot produced thirty sons before producing a single daughter. In the past fanciers claimed that the pigeon's two-egg clutch invariably consisted of a male and female offspring. In a system where sex is determined at random, one half of all pigeon clutches do consist of a male and a female, the other consisting of either two males or two females, but overall the sex ratio is equal and there is no evidence whatsoever that pigeon clutches always or even usually comprise one male and one female chick. If they did it would imply that females had some way of regulating or controlling the sex of each egg they laid. Until recently this was considered impossible and indeed the poultry industry, with its focus on egg-laying females, has endeavoured to find a way of manipulating the sex ratio of chickens for over a hundred years, but without success. However, success now beckons as cloning technology improves and I suspect within a few years there will be one-sex chickens. The ability to do this involves sophisticated modern technology – what is perhaps surprising is that a few species of birds appear to be able to do it by themselves.

Early claims that certain wild birds might be able to adjust the number of male and female offspring were viewed with scepticism, and with good cause. Some of these studies involved relatively small numbers of birds, increasing the possibility of getting a sex ratio different from 50:50 just by chance. Another problem was that the sex of young birds could usually be established only once the chicks had reached a certain age, so it was always possible that the sex ratio is adjusted not by the female herself, but because the chicks of one sex are more likely to die than the other. Molecular methods developed during the 1990s enabled researchers to establish the sex of chicks at the time of hatching, and therefore resolved this particular problem, and in doing so created many new opportunities in sex-ratio research.

The Seychelles warbler, the archetypal 'little brown job' or nondescript small bird, provides one of the most dramatic cases of sex-ratio adjustment in any bird. The species is a cooperative breeder, which means that offspring from previous breeding events, known as helpers, assist breeding pairs, usually by helping to feed the young. It is no coincidence that it should be a cooperatively breeding bird in which a shift in the sex ratio should be so dramatically exhibited. Because helpers are usually daughters from previous breeding attempts (sons generally disperse away from their natal territory) in this species, there is often a very clear benefit to the parents in adjusting the sex ratio of their offspring in favour of daughters. Until 1988 the Seychelles warbler occurred only on the tiny island of Cousin in the Indian Ocean. Following careful management the population of this highly endangered species started to increase and soon all the available habitat on the island – both good and bad – was occupied. For those birds breeding in a high-quality habitat, the helpers made a significant contribution by feeding the young, and breeding females accordingly biased the sex ratio of their single-egg clutch so that around 80 per cent of eggs produce female offspring. In the poorer quality habitat, however, helpers hinder the breeders because they consume an already scarce

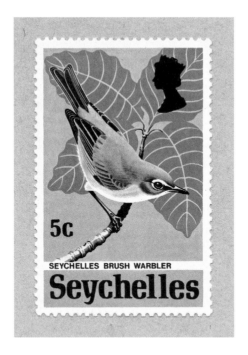

This nondescript little bird, the Seychelles warbler, has two claims to fame: avoiding extinction (through a successful conservation programme), and an unprecedented ability to control the sex of its offspring. Its appearance on a Seychelles postage stamp from the 1972 celebrates only the former.

food supply, and under these circumstances breeding females bias the sex ratio of their offspring to produce sons, which leave home (to look for a territory of their own) and are therefore less of a burden.[33]

How do they do it? Or more to the point, *when* do they do it? If sex determination is adaptive, the timing is crucial. If the sex of an ovum is determined weeks or months before egg-laying, then it is unlikely that a female could ever adjust the sex of her eggs to respond to the current territory quality (or some other cue). On the other hand, if the sex of an ovum is determined just before laying, then it is much more likely that females can adjust the sex of their offspring in an adaptive and strategic manner. And that is exactly what happens. The sex of each ovum is established just a few hours before it is released from the ovary and therefore just before it is fertilised. Most birds ovulate early in the morning, so the circumstances prevailing during the previous day could provide the female with the cues necessary for her to decide whether to produce a male or female egg.[34]

Making sense of sexual anomalies like hermaphroditism, sex change and virgin birth requires us to look at the origin of sex itself. As we have seen, an embryo develops into a male or a female according to its sex chromosomes. The question is how exactly the sex chromosomes transform an embryo into a male or female individual. It is easiest if we start with mammals since they are much better known in this respect. The mammalian Y chromosome carries a gene known as SRY (Sex determining Region on chromosome Y), responsible for transforming what are initially sexless germ cells in the embryo into a testis. Once the testis starts to form it produces its own sex hormones that allow the embryo to continue to develop as male. An ovum fertilised by an X-bearing sperm ends up with two X chromosomes and no SRY gene, and as a result develops as a female. Female mammals therefore develop in the absence of any involvement from the SRY gene – and are therefore considered the default sex.

The situation in birds is different. First, as we saw earlier, the female bird is the heterogametic sex, possessing both a Z and a W sex chromosome, and as a result sex is determined several hours prior to ovulation (whereas in mammals, it is determined at fertilisation). Second, although it would be very convenient if there were a gene similar to the mammalian SRY gene on the W chromosome in female birds, at present we simply do not know whether this is the case. Two other processes could potentially account for which sex an embryo becomes. The first is that a double dose of the Z chromosome in male birds results in the production of a testis. The other possibility is that sex in birds is determined by the ratio of sex chromosomes to other chromosomes (the autosomes): at present we do not know.

Whatever the mechanism of sex determination is in birds, it triggers a cascade of events that culminate in the developing embryo becoming a male or a female. The first cells (known as germ cells) that eventually give rise to the gonads appear in the embryo after just a few days of development and are initially of indeterminate sex and technically known as 'indifferent', meaning that they have the potential to develop as male or female – regardless of whatever sex chromosomes the embryo has inherited. In some ways this potential for the embryonic gonads to become male or female is not too surprising since the testes and ovaries perform identical roles of producing sex cells (sperm or ova) and sex hormones. In mammals it is the SRY gene that launches the cascade of events that turn an indifferent embryo into a male, and the absence of this gene makes the embryo female. It remains to be discovered what the equivalent trigger is in birds that turns the indifferent embryo into a female.

What is known is that in both birds and mammals the homogametic sex is the default sex. Once the gonads are differentiated – that is, once they have started to become male or female – they begin to produce sex hormones. In mammals, testosterone from the testes determines maleness, and in the female an absence of testosterone determines femaleness. In birds, as we might now expect, the opposite is true: oestrogen from the ovary of the female embryo dictates femaleness, and an absence of oestrogen dictates maleness. In other words in the *absence* of any effect of oestrogen a bird embryo will become male by default. In the presence of oestrogen it becomes female.[35]

In both mammals and birds the effects of the sex hormones early in development can be permanent, but if they occur later in development, or even in adulthood, they can be temporary.[36] Sex differences – in appearance, behaviour and brain structure, for example – are therefore determined by two processes, the first of which results in a permanent change and occurs while the individual is still an embryo, and the second, which is much more flexible, occurs later. Under normal circumstances the two processes work in concert, the one reinforcing the other. Occasionally, however, as we have seen, something goes wrong and the two processes become divorced, resulting in various kinds of reproductive anomalies including hermaphroditism, sex change and virgin birth.

Lateral gynandromorphs appear to be the result of an error in an early cell division of the fertilised egg, resulting in one half of the embryo becoming male and the other female, but the exact cause is unknown. An early idea to explain hermaphroditism in mammals was that two sperm (an X and a Y) enter and fuse with the ovum, but this obviously cannot account for gynandromorph birds, because, as we have seen, birds don't have sex-determining sperm. One possibility in birds is that, because of a faulty cell division during its formation, the ovum ends up with two nuclei (Z and W) instead of just one, each of which is then fertilised by a different sperm (both Z) and the error is then maintained in the two halves of the bird as it develops. It sounds like a long shot, but so far it is the most likely explanation.[37]

We have a much better understanding of why birds change sex. Females only ever change into males; the reverse has rarely been reported, and, when it has, it has never been verified. To put this more precisely, genetic females (ZW) only ever adopt the appearance and behaviour of males. This is exactly what we expect if the sexual behaviour and appearance of males is determined by an *absence* of oestrogen. A female bird with a damaged ovary that ceases to produce oestrogen reverts to the default option and becomes male in appearance and behaviour. This effect is not ubiquitous in all birds; a few species, like the ruff, clearly have a different system, as do those species like the house sparrow that exhibit permanent (rather than seasonal) sexual dimorphism.

Finally, virgin birth. It is now clear that an ovum, the female gamete, contains all the material necessary to produce a new individual with no help from a male or his sperm. Since all parthenogenetic offspring are male and must therefore possess at least one Z chromosome, and since they are known to possess a duplicate set of chromosomes, they must be ZZ. Presumably WW offspring are simply not viable.

Science has brought us a long way, successfully resolving many of the mysteries that once surrounded sex. In the past, things we didn't understand were often frightening and if, like an egg-laying cockerel, they seemed to undermine fundamental beliefs they were often dealt with in a barbaric way. In the Middle Ages superstition and a fear of God ruled over logic and common sense. Later, as scientific ideas gained precedence, we were able to make more logical decisions, dispelling fear and replacing it with understanding.

With no dramatic seasonal change in its appearance, the male house sparrow's
(front) plumage is largely under genetic control, unlike species such as the peafowl,
in which hormones regulate the huge seasonal changes in male plumage.
(From Meyer's British Birds, 1835–50)

BLUE POUTER.

A product of artificial selection, this ludicrous bird is a male pouter pigeon in full display.
Standing erect with crop fully inflated, such males are irresistible to female pigeons
and pigeon-fanciers alike. This illustration comes from William Tegetmeier's
Pigeons of 1868.

9

Darwin in Denial

Infidelity

On a small area of waste ground outside the Andalusian town of Gaudix half a dozen middle-aged men are preparing for a contest that will determine their social status for weeks to come. Five of the men each carry a male pigeon bearing a colourful design painted on its wings. The sixth man holds a female pigeon decorated, incongruously, by the addition of three long pink feathers to her tail. At a given signal the female is placed on the ground and the men release their males in front of her. The effect of the female upon the males is electric: they launch themselves into a frenzy of courtship, strutting, tail-sweeping and bowing and cooing for all they are worth. In a tight, highly mobile knot of feathers, the males jostle furiously for the female's attention and it is only by their distinctive wing markings that the men can keep track of how each bird is performing. After a while – and it may take up to an hour – the female accepts one of the males and signals her choice by flying back to his loft with him, at which point the owner, himself puffed up with pride, collects his winnings from the other men.

The Gaudix pigeon event is a contemporary version of the ancient sport in which birds were trained to lure others back to their loft. Pliny noted that 'the pigeon is much given to straying. For they have a trick of exchanging blandishments and enticing other pigeons and coming back with a larger company won by intrigue'.[1] This extraordinary pigeon behaviour became a popular sport in Italy during the Middle Ages. The contests were called *triganieri* and involved two or more groups of pigeons flying around together until summoned back to their lofts by their owners' flags or whistles, when any strangers were then taken captive. Friendly contests saw the captured pigeons returned or ransomed; in less friendly events, the captives were killed.

The Moors brought *triganieri* birds to Spain during the Middle Ages, where they became known as *palomas ladronas*, or 'thief pigeons'. The birds were deliberately bred and trained to lure members of the opposite sex back to their home loft. Not everyone appreciated them, and one eleventh-century law from Seville stated, 'The sale of thief pigeons will be absolutely prohibited, a custom used exclusively by people without religion ...'[2]

The behaviour of the thief pigeons and their 'victims' is absolutely remarkable. Male thief pigeons excel in attracting and accumulating females, copulating with them and abandoning them to rear the offspring alone. An exceptional male can even induce an incubating female to abandon both her partner and her eggs – almost unheard of among wild birds – and return with him to his loft. The fact that females will desert their nest, eggs and partner for a male she deems more attractive flies in the face of just about everything we expect of birds, and certainly of pigeons, which were once considered models of monogamy.

Darwin was well aware of the existence of thief pigeons and also knew from observations of his own birds that female pigeons were occasionally unfaithful. But he was in denial: acknowledging that promiscuity among males was natural, he seems to have ignored the fact that it takes two to tango, and that females were also promiscuous.[3]

Perhaps he was seduced by the long-standing, popular belief in the fidelity of female doves, enunciated (somewhat contradictorily) since Pliny's time:

Next to the partridge, it is in the pigeon that similar tendencies are to be seen in the same respect: but then, chastity is especially observed by it, and promiscuous intercourse is a thing quite unknown. Although inhabiting a domicile in common with others, they will none of them violate the laws of conjugal fidelity: not one will desert its nest, unless it is either widower or widow.[4]

Despite recognising the pigeon's tendency to 'stray', Pliny and everyone else persisted in promoting the pigeon as a paragon of sexual fidelity. For Darwin, male promiscuity was the norm and an important part of his vision of sexual selection. Female promiscuity on the other hand was explicitly *not* part of sexual selection; in Darwin's book females were coy, cautious and, above all, faithful:

It is shown by various facts, given hereafter, and by the results fairly attributable to sexual selection, that the female, though comparatively passive, generally exerts some choice and accepts one male in preference to the others.[5]

The strange thing about this statement is that Darwin knew it wasn't true. He simply chose to ignore all the evidence that females could be promiscuous.

For example, Aristotle had commented on the fact that female fowl sometimes copulated with more than one male, and when they did the offspring resembled the second of the two males: 'and the eggs previously fertilised by another breed of male change their nature to that of the male which copulates later'.[6]

William Harvey, in his *Disputations on Generation* in the 1600s, said:

There are some species of animals in which one male suffices for many females as in the case of hinds and does, and breeds of cattle. And again there are others in which the females are so passionate with desire that they are scarcely satisfied with many males, as the bitch and the she-wolf, and for this reason prostitutes are called 'she-wolves' because they make their bodies public, and brothels 'lupanaria', in which they offer their wares for sale.[7]

Daniel Girton writing in 1765 warned pigeon-breeders that their strains might be 'adulterated by a false tread', which he said 'an over salacious hen will frequently submit to …' thereby providing clear evidence that doves at least occasionally engaged in extra-pair copulations.[8]

William Smellie wrote of the domestic fowl: 'The dunghill cock and hen, in a natural state, pair. In a domestic state, however, the cock is a jealous tyrant, and the hen a prostitute.'[9]

Even more surprisingly, Darwin himself actually describes a case that exemplifies female promiscuity. The information came from William Darwin Fox, his cousin and chum from their Cambridge undergraduate days. Fox was a clergyman and, fascinated by natural history, kept a menagerie at his home in Cheshire. In 1868 he wrote to Charles to tell him about his geese, and the story subsequently appeared in *Descent*:

> The Rev. W. D. Fox informs me that he possessed at the same time a pair of Chinese geese and a common gander with three geese. The two lots kept quite separate, until the Chinese gander seduced one of the common geese to live with him. Moreover, of the young birds hatched from the eggs of the common geese, only four were pure, the other eighteen proving hybrids; so that the Chinese gander seems to have had prepotent charms over the common gander.[10]

So why, in the face of so much evidence, did Darwin deny female promiscuity? There are several possibilities. The first is that he simply did not see it. Darwin's purpose in recounting his cousin's goose story was evidence for female choice, not promiscuity. It also suited Darwin to assume females to be monogamous. Victorian gentlemen did not discuss infidelity – at least, not openly. Assuming females to be monogamous also avoided any embarrassment at home. Darwin's daughter Etty, then in her late twenties, was helping him correct the proofs of *Descent* and so he was extremely careful about what he wrote.[11]

Producing pigeons that will steal the females of other breeders has been a popular sport for centuries and was brought to Europe by the Arabs. Darwin knew of thief pigeons but failed to recognise their significance for his theories. This is an eighteenth-century image of Persian pigeon-keeping from an Arab treatise.

Darwin wasn't alone. Other natural history writers found it politically (or commercially) convenient to assume female fidelity. In some wonderfully florid prose, Buffon (probably disingenuously) extolled the pigeon's social virtues:

> They are fond of society, attached to their companions, and faithful to their mates; a neatness, and still more the art of acquiring the graces, bespeak the desire to please; those tender caresses, those gentle movements, those timid kisses which grow close and rapturous in the moment of bliss; that delicious moment soon renewed by the return of the same appetites, and by the gradual swell of the soothing melting passion; a flame always constant, and ardour continually durable; an undiminished vigour for enjoyment; no caprice, no disgust, no quarrel to disturb the domestic harmony, their whole time devoted to love and progeny; the laborious duties mutually shared; the male assisting his mate in hatching and guarding the young: – If man would copy, what models for imitation.[12]

William Smellie, on the other hand, while acknowledging female promiscuity excuses it as an artefact of domestication. He assumes logically that promiscuity in the domestic fowl is a consequence of their pampered life in captivity, drawing a parallel with those human cultures where abundant resources in the hands of a few powerful males results in them acquiring and jealously defending harems of submissive females.[13]

The idea that birds and other animals might provide appropriate models for human behaviour wasn't new, but it led to a distortion of the facts and there was precious little consistency in whether a particular species exemplified virtue or vice. Pigeon-fancier John Moore, for example, proposed that we should emulate the English Pouter pigeon:

> Separate the old ones, placing them in different coops, and feeding them high with hemp ... then turning them together; and by being

very hearty and salacious, they breed pigeons with very good properties; from whence we may observe, that would mankind be alike abstemious, their progeny might be more complete both in body and mind.[14]

In other words, Moore implies that sexual abstinence somehow results in better quality offspring.

The idea that birds provided moral guidance encouraged religious societies to foster bird-keeping and ornithology, and explains why Anglican organisations like the Society for the Promotion of Christian Knowledge (SPCK) and the Religious Tract Society published so many books on natural history and bird-keeping during the Victorian period. These included Anne Pratt's *Our Native Songsters* and *Our Domestic Fowls and Song Birds* by William Linnaeus Martin who wrote: 'In the creation of animals, whether quadrupeds or birds, expressly serviceable to man, and so highly conducive to his prosperity, and, at the same time, so easily subjugated or tamed, we cannot but see the wisdom and goodness of Divine Providence.'[15] God's creatures also provided role models for human behaviour and the British ornithologist the Reverend Frederick Morris advocated (mistakenly, as we'll see) that his parishioners emulate the dunnock:

Unobtrusive, quiet, and retiring, without being shy, humble and homely in its deportment and habits, sober and unpretending in its dress, while still neat and graceful, the Dunnock exhibits a pattern which many of a higher grade might imitate, with advantage to themselves and benefit to others through an improved example.[16]

Darwin's influence on biology was such that his statement about female fidelity resulted in a century of mistaken belief about their sexual behaviour. Even in the 1950s and 1960s, when field studies of the behaviour of individually marked birds were just starting to yield rich returns, ornithologists were still in denial. When they saw extra-pair matings in species thought to be monogamous, they simply discounted the possibility that females were in any way responsible, and sometimes even excused the males by suggesting they were sick or suffering from a hormone imbalance.[17] In other words, as recently as fifty years ago infidelity among socially monogamous birds was considered a mistake.

The change occurred in the late 1960s, as ideas about the way evolution occurred were brought into sharper focus. Over the preceding years, biologists had allowed the idea of natural selection to drift away from Darwin's original, ruthless process that cut down or promoted individuals, and instead – in certain quarters at least, and especially in the public's eye – had become more benign and more concerned with preserving the species. With this mindset it is easy to see how ornithologists prior to the 1960s might have struggled to make sense of infidelity: what possible advantage could there be *to the species* of one male inseminating another's partner?

Not everyone at that time thought of natural selection operating on species. David Lack for one did not. He and a handful of others stuck rigorously to Darwin's original view that selection worked on individuals. Most influential of these was George Williams, then at Stony Brook University, New York, whose book *Adaptation and Natural Selection* was written to counter such species-centered evolution. It launched a new era in evolutionary thinking and in bird biology in particular.[18]

Two young biologists, Geoff Parker and Bob Trivers, neither of whom were strictly ornithologists, were keen advocates of Williams' new way of thinking. Parker, an amateur breeder of exhibition poultry and a professional zoologist, studied the promiscuous mating behaviour of dungflies. Bob Trivers, more interested in ideas than animals, was none the less inspired by the behaviour of the pigeons on his Harvard apartment windowsill:

What soon became clear in this monogamous species was that males were sexually much more insecure than were females, and males acted to deprive their mates of what they would be happy to indulge in themselves, that is, an extra-pair copulation ... the group outside my window began with four pigeons – two mated couples. They slept next to each other in the gutter of the roof of the house next door ... the two males, although they were the more aggressive sex, always sat next to each other with each one's mate on the outside. By sitting next to each other, the males could ensure that each one was sitting between his mate and the other male.[19]

More or less independently, Trivers and Parker capitalised on individual selection to create a new vision of sexual selection. With an explicitly individual perspective reproduction was no longer the cosy, cooperative affair between male and female operating for the good of the species, but instead was a battleground in which members of each sex competed among themselves, attempting to exploit the other in a struggle for genetic representation. Sexual selection was about producing descendants and, from an evolutionary perspective, genetic descendants were all that mattered.

Thanks to Geoff Parker and Bob Trivers sexual selection was reborn in the late 1960s. What they said was what Darwin had meant all along, although, to be fair, he hadn't always said it as clearly as he might have. Parker and Trivers picked up this new version of evolution and ran with it and so rich were the insights it provided that biologists have been running with it ever since.

By assuming females to be sexually monogamous Darwin automatically imposed the view that sexual selection came to an end once an individual of either sex had acquired a mating partner. What Parker recognised as he watched his female dungflies copulate with a succession of males was that sexual selection also continues *after* insemination and carries on, right up to the moment of fertilisation. Males compete for fertilisations, not females.[20]

By being promiscuous a female carries the sperm of several males simultaneously and those sperm then compete to fertilise her eggs. Sperm competition, as Parker called it, is a potent evolutionary force, for in an evolutionary sense it pays a male to protect his own paternity, but at the same time it also pays to steal paternity from other males.

As Parker recognised, the conflict this inevitably creates in the evolutionary struggle for descendants drives males to outdo each other, through their behaviour, anatomy or physiology – in any way possible. For the time being, females were not considered.

Acknowledging the existence of sperm competition allowed other pieces of the reproductive puzzle to fall into place. For centuries, biologists and others had pondered the significance of unusual reproductive phenomena. This new view of sexual selection – it was called 'post-copulatory' to focus attention on the fact that the battle for paternity continued after copulation and insemination – provided answers to many questions, including one asked by John Ray in *The Wisdom of God*:

> Why should there be implanted in each sex such a vehement and inexpugnable appetite of copulation?[21]

Here, Ray may have been thinking of the house sparrow, whose copulatory prowess was legendary:

> Its desire for coitus and reproduction is so compelling that it may copulate as many as twenty times an hour.[22]

In 1559 Dresden's parson Daniel Greysser was commended for his Christian zeal in having 'put under ban the sparrows, on account of their unceasing and extremely vexatious chatterings and scandalous unchastity during the sermon'.[23]

We now know the reason for their inexpugnable appetite: female promiscuity. If females are promiscuous a male's best chance of being the true father of the chicks he helps to rear is to copulate repeatedly with his partner. House sparrows, for example, form season-long pair bonds, but infidelity is common and some 10 to 15 per cent of all chicks are the outcome of extra-pair copulations. The figure would be higher still if male partners did not copulate so frequently with their female. The rules are pretty much the same for other birds, too,

Despite appearances, the majority of small birds are not sexually monogamous and most broods, like these European blackbirds, contain one or two illegitimate offspring. (From Travies, 1857)

and indeed elsewhere in the animal kingdom; frequent copulation is a paternity assurance device and therefore an evolutionary imperative. I cannot help feeling that, had this been explained to Ray, he would have seen its compelling logic.

For a long time the actual mechanics of copulation and insemination in birds was a mystery. It was made more confusing by the fact that the males of only certain birds such as swans, ducks and ostriches had, like mammals, a penis, whereas others like the fowl and small birds did not.

Albert the Great provides one of the first accounts of copulation in birds, describing what happens in swans, but apparently without actually catching sight of the male's curiously coiled penis:

> At mating time, the males lean their necks on the females by way of a caress. Then the male ascends the female and projects his semen into her. It is said that the female receives the semen with pain and that therefore, after copulation, she flees from the male. This is false because he put in her only that humor which is received with pleasure. She flees because at this time the desire for copulation ceases. After copulation, moreover, both the male and the female dip themselves in the water, as do aquatic birds. This is because the vapor of their concupiscence [strong desire] is running through their flesh and they are seeking to purge it. A proof of this is that all birds have bristling feathers after copulation and they shake themselves with feathers raised and spread.[24]

William Harvey also made very detailed observations of copulation in his frustrated quest to understand the details of reproduction. With privileged access to the royal menageries, Harvey was able to witness at close range the extraordinary copulation of ostriches:

> I myself have seen a hen ostrich, when her keeper gently stroked her back with the intent to arouse her desire, throw herself on the ground, lift aside that veil and disclose and stretch out her vulva. When the

cock bird saw this, being instantly inflamed with desire, he mounted her, and with one foot on the ground and the other pressing on her back as she lay, accomplished his purpose with an exceedingly large and vibrant yard [penis] that you might have taken for a neat's [a kind of cow] tongue. All this went on with much muttering and noise on both their parts, stretching out and pulling back their heads and many other signs of rejoicing.[25]

Harvey dissected the reproductive organs of ostriches that died in the menagerie, rejoicing in the birds' enormous size that helped him figure out what was what.

In a cock ostrich I found ... a very large glans and a reddish yard ... and I have often seen it rigid in condition and vibrating and somewhat embowed. But when he had inserted into the vulva of the hen he kept it there for a long time without any movement at all, as if in coition they had been fixed together by some stake.

In a little aside Harvey also tells us:

In a black drake I once saw a penis of such length that, after coition, a hen pursued it as it trailed along the ground and eagerly pecked at it, believing it, I am sure, to have been a worm, and this made the drake retract it more quickly than is his custom.[26]

After what he had witnessed in the ostrich, drake, cob-swan and gander, Harvey was puzzled by the absence of a penis in the rooster: 'In its place I find in the cock an orifice, just as in the hen, but smaller and narrower ... I think the same thing happens in the cock, who has no penis, as in smaller birds who perform the act of coition quickly and only by rubbing.' An astute observer, Harvey is absolutely correct: most small birds copulate very briefly, usually for just one or two seconds. In both sexes the opening of the reproductive tract lies inside the cloaca (next to the opening of the gut) and copulation comprises

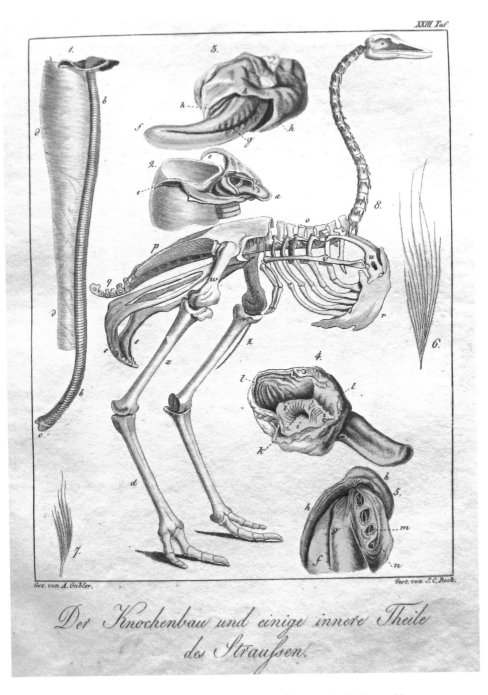

Ostrich anatomy: William Harvey described the ostrich phallus as 'like a cow's tongue' – as is apparent in this illustration. (From Wolf, 1818)

the coming together of the male and female cloaca in what has been referred to euphemistically as a 'cloacal kiss'.

Although small birds do not possess a penis, the males of some species do have a structure that superficially, at least, appears similar. The 'cloacal protuberance', as it is called, is formed mainly from the paired, highly coiled distal ends of the vas deferens (referred to as the seminal glomera) that contain the males' sperm store. In birds and most mammals, sperm survive best when stored below body temperature and because the cloacal protuberance projects from the body, the sperm it contains are 4 or 5°C cooler than they would otherwise be.[27] The equivalent structure in humans (and other mammals) is the epididymis, which, instead of being separated from the testes as it is in birds, is wrapped around the outside of each testis, which in turn are housed outside the body cavity in the scrotum, thereby ensuring that their precious cargo is kept cool.

John Hunter made a comparison of the gonads of both mammals and birds in the late 1700s, dissecting house sparrows, ducks and geese and by the ingenious method of injecting mercury into one end of the human epididymis (of a cadaver) showed that it comprised a single convoluted tube connecting the testes to the vas deferens. Similarly, he realised that in the sparrow the enlarged end of the vas deferens was a sperm store and hence an essential requisite for their frequent copulation.[28]

In some birds the male's cloacal protuberance is so large during the breeding season that it can be used to identify the sex of species that are otherwise indistinguishable. When Edwin Mason 'discovered' this in the 1930s, he said that while his findings might not be new, 'no reference to the ability of being able to determine the sex in living birds except by plumage and measurements had ever come to the writer's attention ...' In fact, as he suspected, the method had been known for centuries, at least to bird-catchers:

As far as those [nightingales] which one takes in March are concerned, the identification is not only laid in song, but also in the lower sexual

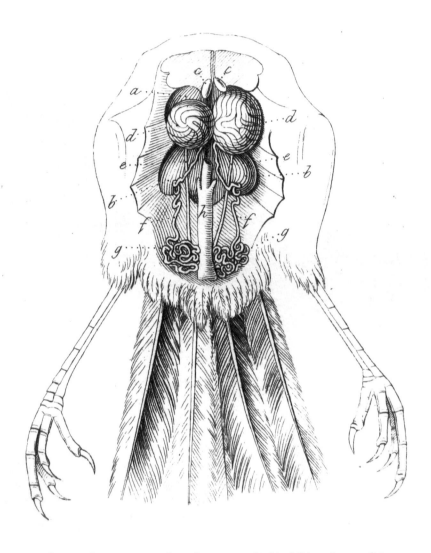

The reproductive system of a male passerine bird in full breeding condition, showing the testes (d), vas deferens (f) and seminal glomera (g), which form the cloacal protuberance. (From Wagner, 1841)

parts, which the males throw outward, in contrast with the females, because that is the time when birds mate. Those are thus the very evident signs and arguments, to which you keep yourself.[29]

I asked nightingale researcher Valentin Amrhein of the University of Basel about this and he confirmed that while the male nightingale's cloacal protuberance is not as large as in some other species, at the time of pair formation the difference between the sexes is obvious.[30]

Most eighteenth- and nineteenth-century books on bird-keeping include a section entitled 'To know the cock from the hen', almost certainly because people were more interested in owning males – because of their song – than females. Surprisingly, there are only a few references to sexing birds from the cloacal protuberance in these accounts. Canaries must have been among the most commonly kept birds in the eighteenth century and the sexes are difficult to distinguish on plumage alone, yet there are few references to the cloacal protuberance as a way of telling the sexes apart.[31] To anyone who has kept canaries, the male's protuberance is very obvious during the breeding season. The great canary expert of the early 1700s, M. Hervieux, alludes to it thus: 'when the male is in love, then one can make out the button which is longer than that of the female'.[32]

The failure of most writers to mention the male's cloacal protuberance is puzzling, especially since in some songbird species, like the dunnock, it is quite pronounced. Notwithstanding the Reverend Morris's encouragement for his parishioners to emulate its habits, the dunnock is actually among the most promiscuous of birds, copulating dozens of time each day, necessitating a large protuberance.[33]

The dunnock's close relative, the Alpine accentor, has an even more promiscuous mating system and an even larger cloacal protuberance, first described by a Swiss ornithologist who, overwhelmed by what he saw, wrote:

Overleaf: *Among the most promiscuous of birds, the Alpine accentor's modest plumage belies its extraordinary sex life, as well as obscuring parts of the male's remarkable reproductive anatomy. (From the Reverend F. O. Morris's* A History of British Birds, *1856)*

The vas deferens of the Accentor, instead of opening directly in the cloaca, meander at its walls in two large compact clews; these seminiferous (seed-carrying) balls, almost egg-shaped, wrapped in a 'leaf' [skin] derived from the peritoneal membrane, hang like sacks, at both sides of the anus, in pockets formed by skin of the body and supported by the pubic bone. And finally: Do the symmetric balls that hang, in spring, beneath the tail of the Accentor not a little remind one, by their position, of testicles of the superior animals?[34]

They do indeed, and so much so that when Fatio's description was later used in the monumental handbook *The Birds of the Western Palearctic* it became wonderfully garbled: 'Extreme rapidity of copulation perhaps facilitated by male's greatly elongated and sinuous sperm duct which hangs down as egg-shaped sac ... either side of cloacal aperture ...' The confusing image these words conjure up reflects how unfamiliar most ornithologists were with the male reproductive system, even in the 1980s. In fact the Alpine accentor's extraordinary attributes made sense only when it was discovered what a prominent role sperm competition plays in this species' life – its cloacal protuberance was simply a much enlarged version of what occurs in the dunnock and many other species.[35]

At the other end of the scale there are some species where the male has virtually no cloacal protuberance. In 2000 I was told by Sean Fitzpatrick, a Yorkshire bird-breeder, that the cloacal protuberance of the male bullfinch was tiny and the same size as that of the female. This suggested to me that this species' sperm stores and perhaps its testes, too, would be relatively small, as I was later able to confirm, indicating that sperm competition is probably absent in this species.[36]

During their continental tour in the 1660s, Francis Willughby and John Ray routinely dissected birds they obtained from bird-catchers and in city markets. Other than shooting them, there was no other way to study birds. Prior to setting off Ray had been in the habit of dissecting birds to see what they had been eating, but now he and Willughby systematically recorded the size and form of the internal organs, including the testes of every bird they could get their hands on. The fact that in birds the testes are internal, rather than external as in humans and most other mammals, was first noted – inevitably – by Aristotle and, much later, commented on by William Harvey.[37]

The more subtle observation that the *relative* size of the testes differed between bird species was also first recorded by Aristotle, but it was Ray and Willughby who recognised the biological significance of this fact. Referring to the quail they say: 'The cock has great testicles for the bigness of its body, whence we may infer that it is a salacious bird'; similarly (and with uncanny accuracy), for the dunnock they noted that: 'The cock [has] large testicles', and of the house sparrow: 'Its testicles are great, as being a very salacious bird.'[38] Building on these observations perhaps, Buffon says: 'The bulk of these kinds of glands [testes] is far from being proportioned to the size of the bird. In the eagle, they are only the size of a pea; in a cock of four months old, they are as large as olives.' Ray and Willughby were absolutely right in assuming that relatively large testes were linked with 'salaciousness'.

In the 1970s Roger Short, who was at the Unit of Reproductive Biology in Edinburgh, made measurements of the testicles of (anaesthetised) primates. On finding a huge discrepancy in the absolute and relative size of the testicles of the chimpanzee and gorilla, Short suggested that the chimpanzee's enormous testicles had evolved because the males copulated so frequently. Soon afterwards Short learnt about Geoff Parker's research and realised that sperm competition provided an even more compelling explanation for the difference in testes size between chimps and gorillas.[39] He was right, and we now know that across a wide range of animal groups, from insects, fish, frogs, mammals and birds, relatively large testes are – strange as it may seem – the hallmark of female promiscuity. Here's the logic: in species in which females are sexually promiscuous males have evolved relatively larger testes, because larger testes produce more sperm, and the more sperm a male transfers to a promiscuous female the more likely it is that he – rather than any of her other copulation partners – will fertilise her eggs. In terms of getting genes into the next generation, large testes give the males of species such as the dunnock, Alpine accentor and quail, whose females are promiscuous, an evolutionary advantage.[40]

The testes of birds – unlike those of most mammals – also show enormous seasonal variation in size, a fact noted first by Aristotle and elaborated upon by Albert the Great:

> Sometimes, these testicles are not visible during a time of sexual inactivity, but they swell and are visible when they swell with semen during the breeding season: Likewise, before the laying season, the testicles of birds are small and escape notice, but they enlarge during the mating season. This is especially noticeable in turtledoves and partridges. This is why many have held that turtledoves and partridges and certain other birds have no testicles in winter.[41]

Strikingly, Albert also noted that birds which copulated frequently had relatively larger testes:

> It is a common trait of birds that their testicles grow during the mating time. The testicles of the ones that have the most copulations grow the most and they are then found to have the largest testicles, such as the cubech [partridge] and chickens. When, however, they do not copulate, they have small testicles.[42]

Several later authors, including Buffon, Hunter, Ray and Willughby, also commented on the seasonal change in testes size, but it was the ever-imaginative John Hunter who first made systematic observations and measurements. One winter in the late 1760s he caught six male house sparrows and kept them in his menagerie, sacrificing them in turn as the breeding season progressed: 'If we compare their [the testes'] size in January with what it is in April, it hardly appears possible that such a wonderful change could take place during so short a period.' In Hunter's original publication the variously sized testes from the six birds are illustrated in glorious but unhelpful isolation, but the sparrows with their testes intact are still on display in the Hunter Museum in London.[43]

In writing his account of the ruff in his *Natural History of Birds* Buffon suggests that the possession of both ornamental feathers and the 'eruption of fleshy turgid pimples on the fore-side of the head and round the eyes ... implies great prolific powers'.[44] As though to confirm this he refers to information on the ruff sent to him by Louis Baillon, who says: 'I know of no bird in which the appetite of love is more ardent; none whose testicles are so large in proportion.'[45] However, Baillon's observation on the relatively enormous testes of the ruff was completely overlooked by subsequent ornithologists. When the Danish ornithologist Anders Møller undertook his pioneering comparative study of testes size in birds in the early 1990s, his results indicated that birds with a lek mating system (and this included the ruff, although there was no information on the size of the ruff's testes) all have relatively small testes. His explanation was that sperm competition is unlikely among lekking species because females are completely free to choose their copulation partner and therefore have no need to engage in copulations with more than one male and males can therefore afford to have tiny testes.[46] For most lekking species this turned out to be true – with the exception of the ruff. Detailed studies later showed that sperm competition *does* occur in ruffs and that the level of multiple paternity is surprisingly high.[47] Knowing of my interest in reproduction, a researcher at Uppsala University who was studying ruffs in the 1990s shot some males on a lek and sent them to me to dissect. I was amazed by the species' enormous testes, confirming Baillon's description – which I didn't find for another decade, a clear example of why history should not be ignored.

Previous generations of ornithologists are equally guilty of overlooking opportunities and John Ray, for example, failed to realise the insights a microscope might provide. When Leeuwenhoek described his own sperm in the late 1600s, he also commented on how much they differed from those of a cockerel. Mammalian sperm were like tadpoles, with a paddle-shaped head, whereas those of the cockerel were pointed and more snake-like in appearance. Much later, in the 1830s, the German zoologist Rudolf Wagner also noticed the marked difference in sperm design that exists between passerine and non-passerine birds. A pupil of Cuvier's, Wagner was among the first to make use of the new compound microscopes developed in the 1830s[48] and had this to say about the sperm of birds:

> [They] have a long-shaped body with thread-shaped appendix. Those of the passerines ... are without exception distinguished by their very long, straight flagellum, [and] corkscrew-shaped, spiral end.[49]

Later, the Swedish biologist Gustaf Retzius, whose interests spanned poetry, anthropology and reproduction, confirmed the fundamental difference in sperm design between passerines and non-passerines, but also showed just how great the variation in sperm design was across different bird species within each group. His drawings of sperm are both beautiful and extremely accurate, establishing him among the cognoscenti as the 'father of comparative spermatology'.[50]

In a world in which females are at least occasionally promiscuous, keeping other males' sperm at bay is a fundamental issue for males. Trivers' observations of male pigeons jostling for position and keeping others at a safe distance is now referred to as mate-guarding, an adaptation to protect a male's paternity. Several early ornithologists such as George Montagu witnessed male birds following their female during the time she was nest-building and egg-laying, but had no framework with which to interpret it, and no idea that this was mate-guarding:

> When the female [nightingale] has chosen a spot for nidification [nesting], the male constantly attends her flight to and from the place, and sits upon some branch near, while his mate instinctively places the small portion of material she each time brings to rear a commodius fabric for her intended brood.[51]

A century later Edmund Selous made almost identical observations on European blackbirds: 'the cock is as busy in escorting and observing

The male yellowhammer remains close to his partner throughout the several days that she can be fertilised. This illustration is by George Lodge (from Howard, 1929). The upper image shows the anterior part of a yellowhammer's sperm (from Retzius, 1909).

the hen as she is in collecting material for the nest', but again he offered no explanation. And Eliot Howard says something very similar for almost every one of the different warbler species he watched.[52] The significance of this following behaviour, which we now know is almost ubiquitous in small birds, was not recognised until the shift in evolutionary thinking in the late 1970s. It made sense only once we recognised that females are not sexually monogamous, and that the period over which females can be fertilised spans several days. As predicted, there exists an extremely close match between this fertile period and the male partner's obsession with keeping close to his female.[53]

One of the most striking differences in reproductive behaviour between birds and other animals is that most birds very obviously breed in pairs. William Smellie summarised much of what was known about the mating systems of birds in the eighteenth century by saying:

Marriage or pairing, though by no means a universal institution of nature, is not unfrequently exhibited in the animal creation. With regard to man, both male and female are instinctively impelled to make a selection ... This universal, and almost irresistible impulse of selection, is to me the strongest argument in favour of monogamy, or the union of pairs, among the human species ... The same impulse, or law of nature takes place among many other animals, as the partridge tribes, the swallow, the linnet, and, in general, all the small birds. The assiduity, attention, mutual affection, laborious vigilance, and steadfast fidelity of pairing animals, are truly admirable, and, to ingenuous minds, afford the most exemplary admonitions to virtue and conjugal attachment.[54]

Smellie also anticipates the evolutionary thinking of Darwin, and especially David Lack two centuries later, when he says:

All those species of animals, whose offspring require, for some time, the industry and support of both parents, are endowed with the instinct of selection, or of pairing. With regard to the feathered tribes, pairing is almost universal ... The mother is not, like quadrupeds, provided with organs fitted to secrete milk ... She is therefore obliged to go abroad in quest of food for them. But the progeny are so numerous, that all her industry, if not assisted by the father, would be ineffectual for their support and protection.

He continues:

There are other species of pairing birds, whose young, as soon as they are hatched, are capable of eating their food ... and of course, require less labour from the parents. In these species, accordingly, the male pays no attention to the progeny, because it is unnecessary ...[55]

In the 1960s David Lack estimated that over 90 per cent of all birds were socially monogamous, an arrangement in which a male and female work together to rear offspring. Other mating systems include

Many birds of paradise, like these crested birds of paradise, have a lek-based mating system. This painting is by Joseph Wolf from Daniel Elliot's monograph on the birds of paradise published in 1873, when this species was known as the golden bird of paradise.

polygyny (literally, many females), where some males have several female partners, as in lekking ruff, peafowl or black-grouse, and where females visit leks simply to copulate. Another form of polygyny involves a single male being paired to several females simultaneously, as in the red-winged blackbird. The rarest mating system of all among birds is polyandry (literally, many males) where a female simultaneously has several male partners, as in jacanas.[56] What Lack did not anticipate was that, although the majority of birds were socially monogamous, they are sexually quite promiscuous.

It is curious how some scientific discoveries seem to occur at exactly the right time. Just as Trivers' and Parker's ideas about sexual selection were becoming established, what ornithology needed more than anything else was a way accurately to measure male reproductive success. Developed in the 1970s, DNA fingerprinting provided for the first time a way of unambiguously assigning paternity. Prior to this, ornithologists could do little more than speculate whether the extra-pair copulations they witnessed resulted in fertilisation. The new molecular tools, coupled with the new theory, overturned two millennia of accepted wisdom about the marital arrangement of birds. As more species were studied it became apparent that *sexually* monogamous birds were the exception rather than the rule: female promiscuity and sperm competition were virtually ubiquitous.

Armed with this new knowledge, ornithologists were suddenly able to make sense of what had previously been mystery. It now became obvious why some species have relatively large testes; why males trail after females during the several days around egg-laying; why some birds copulate so frequently, or for an extraordinarily long time; why some species have a penis – all of these are adaptations to sperm competition.[57] Ray would probably have been shocked, for none of these traits was easily explained in terms of a benign God; they made sense only within the framework of individual selection. Without this, it is hardly surprising that ornithologists did no more than describe what they saw, resisting the temptation to speculate. Of course, Ray did speculate a little, noting that birds with large testes were particularly salacious, but this didn't answer the question of why bird species should differ in their inexpugnable appetite for sex. From the 1970s the new theory of sexual selection provided a framework that allowed ornithologists to answer questions like this for the first time.

This does not mean that we fully understand the mating systems of birds – far from it: there are still many unanswered questions. While males benefit from copulating outside their pair bond by leaving more descendants, the evolutionary benefits of female promiscuity are far less obvious and the question of why females pair with one male only to seek copulations with others continues to vex ornithologists.[58]

My guess is that it will take a major shift in our thinking to resolve this issue: it may happen tomorrow, it may take a decade, or even a century.

Pet parrots, like this African grey, provided some of the first longevity records
for birds. This image is from a painting by Frans Mieris,
Lady in Red Jacket Feeding a Parrot, *dating from the 1660–70s.*

10

A Degenerate Life Corrupts

Reproduction and Longevity

One of the joys of being a professional ornithologist is that it has allowed me to study so many bird species and in all corners of the globe. Each summer for the last thirty-five years I have visited the beautiful island of Skomer off the Welsh coast to study its seabirds. The guillemots, the little penguins of the north, are my main focus on that island but I am equally passionate about the shearwaters, puffins and razorbills that come there to breed. Home to hundreds of thousands of seabirds, Skomer is one of a handful of islands around the British coast where the numbers of seabirds have been counted each year for several decades. Thousands of these birds have also been banded with both metal and colour rings, allowing researchers to estimate how long they live, and ultimately to understand why their numbers might be changing.

I love clambering over the thrift-covered cliffs in search of my colour-ringed birds, whose breeding sites are the rocky ledges high above the Atlantic swells. A few years ago I was catching razorbills with a fibreglass fishing rod tipped with a bent piece of wire, a design based on the traditional sea-fowler's hook, to hook the birds by one of their legs. I was at the eastern end of the island, sliding across the cliff-top on my belly towards a razorbill bearing only a metal ring. If I could just capture this bird it would become a 're-trap' and yield some information about its history. With an open beak, betraying its uncertainty, the bird shuffled away from me and the approaching hook, lifting its wings as though to take flight. Held firm by its attachment to a nearby partner perhaps, it changed its mind. Then, distracted by a gull swooping overhead, the razorbill glanced skywards and as it did so, I slipped the hook around its leg. Snarling in angry disbelief, the razorbill became a formidable and aptly named bird in the hand. Holding its head so it could not bite me, I turned the bird around so that I could read the numbers on the heavily worn metal ring. Placing a colour ring on the other leg so that in future the bird could be identified without being captured, I let it go and watched it fly out to sea.

I discovered that this was a rather special razorbill, having been ringed thirty-seven years earlier, not far from where I had caught it. I later found out that this was the oldest razorbill ever recorded, worldwide.

Thirty-seven seems old, but some species live even longer: the longevity records go to another group of seabirds, the albatrosses and petrels, some of which have been found alive fifty years after first being ringed.

Another species I have studied for many years breeds in the southern hemisphere, in Australia's outback, where the air is hot and redolent with scent of burnt grass and distant eucalypt. The temperature in the shade is 45°C and the only sound is the gentle rustling of the tired leaves in the thorn bushes. In the shade of that vegetation sits a group of tiny birds, with beaks gaping and wings held away from their bodies as they try to capture a little breeze to cool themselves. These are zebra finches, one of Australia's toughest little birds. The females are uniform grey with a reddish beak, but the males sport a patchwork of colour and patterns: a bright red beak, orange cheek patches, barred chest and white-spotted chestnut flanks.

These are extraordinary little birds. Weighing just 12 g (the same as three regular-sized sugar cubes), they are one of several species designed to cope with the climatic extremes of Australia's red interior – hot days, cold nights and drought. Zebra finches can survive without water for weeks, miraculously extracting sufficient moisture from the dry seeds they eat. When it rains, though, their response is remarkable and within a few days they burst into breeding condition, ready to capitalise on the flush of new grass. With enough rain the grass will flower and provide soft, milky seeds on which the birds can rear their chicks. The zebra finch is a tiny breeding machine and, as long as conditions remain good, it will continue to breed, barely pausing between broods.

As the furnace-like heat of the afternoon dies away the zebra finches start to relax, tidying up their plumage and hopping among the branches while calling their miniature-trumpet-like *beep-beep-beep*. With the heat shimmer gone I can use my binoculars again to see the coloured plastic rings that some birds have on their legs. For several weeks I have been watching these birds as the guest of Richard Zann, who has studied zebra finches for much of his working life. The unique combinations of coloured rings allow us to recognise and become familiar with certain individuals; I know their partners, where they nest, how many broods they have produced, and, from Richard's records, how old they are. This last piece of information is startling. All of the birds are just a few months old, but Richard tells me that none will survive longer than twelve months, making these among the shortest-lived birds on earth.

So why should seabirds like the razorbill and the albatross live forty or fifty times as long as a zebra finch? Why should different species of bird have such dramatically different lifespans?

Remarkably, considering he had so little information, Aristotle thought he knew the answer – that too much sex is bad for one's health. The fact that sex, or more generally reproduction, can be both physically exhausting and reduce longevity has been a joke, among men at least, for millennia. Aristotle was convinced that 'salacious animals' and those with 'abundant seed' used up the moisture in their bodies, leaving themselves vulnerable to rapid ageing. Albert the Great claimed that sparrows 'live but a short time due to their [frequent] copulation'.[1]

Throughout history men have been obsessed with their own mortality and Aristotle devoted many pages of the *History of Animals* to how long different species live. Aristotle mentions captive partridges living for fifteen years and a peacock that reached twenty-five years. Pliny reports turtle doves surviving for eight years and, less plausibly, pelicans and eagles reaching eighty and ninety-three respectively. In the tenth century Isidore of Seville made the bold statement that vultures live for one hundred years, but this was obviously a guess, and probably based on an implicit assumption – derived from Aristotle perhaps – that larger animals lived longer than small ones. More reliably, in the sixteenth century Gessner described a 23-year-old captive goldfinch which was obviously geriatric: its plumage had turned grey, its beak had to be trimmed each week, and on being removed from his cage the bird lay perfectly still, unable to fly.[2]

By the sixteenth century a surprising amount of information on the lifespan of captive birds was available. Writing in his bird-keeping book in 1575 Cesare Manzini, for example, provides a comprehensive summary of the longevity of captive songbirds, emphasising not only the variability within and between species, but also the process of senescence itself:

> If one wants to know how long the birds live, understand that amongst the nightingales some live three years, others five, others up to eight years, and sing until that time. Henceforth they are not perfect anymore, thus go deteriorating little by little. Nightingales have been seen which have lived up to fifteen years, and always singing a little or sufficiently, so that they live according to a good treatment which they receive, or according to their constitution.[3]

Two centuries later Buffon had this to say about a captive nightingale:

> I have seen a nightingale which … has survived until its seventeenth year: that old bird had started to become grey since an age of seven

334

years; at fifteen he had entirely white feathers at the wings and tail; his legs, or better thighs, had much fattened, by the extraordinary growth of the scales (*lames*), with which these parts are covered in the birds; finally he had sorts of bumps at the toes as a sufferer of gout, and from time to time the point of the upper mandible had to be clipped, but he did not have the discomforts of old age; he was always cheerful, always singing, as in his best days, always caressing the hand that fed him. It should be remarked that this nightingale had never been paired: love seems to shorten the days.[4]

Manzini gives the maximum ages of a variety of birds – rock-thrush up to five years; goldfinches ten, even fifteen to twenty years; wood, crested and skylarks, linnets and greenfinches up to five years; siskin six years (though he adds that this isn't necessarily desirable since the siskin's song is so irritating). Finally, he reports that Spanish canaries live about ten years, with a few reaching twenty years of age.[5]

The maximum lifespans of these species in captivity reported by Manzini are all reasonably accurate, as later records confirm. His point in recounting them was so that bird-keepers might have some idea how long their pets would survive, but one senses that he had an inherent fascination with mortality and his own vulnerability.

Inevitably, some philosophers questioned whether information on longevity based on captive animals was representative of what happened in nature. Francis Bacon, writing in the 1600s, thought that it was: 'In tame creatures their degenerate life corrupteth them; in wild creatures their exposing to all weathers often intercepteth them.' In other words, whereas birds in captivity may be protected from the vagaries of the weather, they are 'corrupted' by too much sex and too much food; wild birds avoid such problems but have to deal with the rigours of the weather. The costs and benefits cancel each other out, he assumed.[6]

Only a few years later, Ray and Willughby suggested that wild birds were longer-lived: 'there is no doubt but birds that enjoy their liberty, living at large in the open air, and using their natural and proper food, in gathering of which they also exercise their bodies, live much longer than those that are imprisoned in houses and cages'.[7] John Knapp, on the other hand, in his early nineteenth-century *Journal of a Naturalist*, thought that cage-birds would live longer than wild birds: 'the eagle, the raven, the parrot, &c., in a domestic state, attain great longevity; and though we suppose them naturally tenacious of life, yet, in a really wild state, they would probably expire before the period which they attain when under our attention and care'.[8]

Knapp was right since a comparison of the longevity of captive and wild birds by the ecologist Bob Ricklefs in 2000 showed that captive birds lived 30 per cent longer than their wild counterparts. The really important finding from his study, however, was that that the intrinsic pattern of ageing was identical in captive and wild birds – birds that tended to be long-lived in captivity were also those that survived longest in the wild.[9]

The early Greek philosophers had three main ideas to explain why lifespan differed among different animal species. First, Aristotle pointed out that in mammals there is a pronounced association between longevity and the time spent *in utero*. Animals like mice and rats, which spend only a few weeks as a developing embryo inside their mother, have shorter adult lives than species like cattle and elephants where gestation is more protracted.[10] Second, Aristotle suggested that larger birds lived longer lives than small ones. Third, as we have seen, he confidently assumed a link between lifespan and sexual activity, including the notion that excessive copulation left males at least vulnerable to rapid ageing and early death. In addition, he thought that those species vulnerable to predators 'compensate' for their reduced lifespan by producing more offspring; either reproducing more frequently, and/or having larger broods and, as a consequence, dying young.[11]

All three of these early beliefs have some basis in truth, but let's start with Aristotle's first two ideas.

Lodola

These two paintings from 1620–21 of a woodlark (top) and skylark by Vincenzo Leonardi were the basis for the engravings in Ray's Ornithology (1676, 1678). Larks were common cage birds, with a maximum life span in captivity of about five years.

TAB. X.

The cockerel's intemperate lust was thought to enfeeble its body.
This extraordinary image of a plucked cockerel is from George Stubbs (1804–06).

Aristotle's presumed link between longevity and the time spent *in utero* has since been verified by twentieth-century researchers studying mammals.[12] Intriguingly, Ray and Willughby felt that it might also be true in birds in the sense that 'time *in utero*' was equivalent to the duration of incubation: the interval between laying and hatching. But no one appeared ever to have tested this idea. In their recent comprehensive study of the longevity of birds, Peter Bennett and Ian Owens did not consider Aristotle's idea because they felt it unlikely. But when I asked them to look at their database and check, it turned out that Ray and Willughby were correct – birds with longer incubation periods do indeed tend to be longer-lived.[13]

Aristotle's second idea, that large animals live longer than smaller ones, also turns out to be broadly true, as shown by simply comparing the short-lived zebra finch – one of the smallest of birds – to albatrosses, which are among the largest. However, as we shall see, body size on its own is not a particularly good predictor of how long a bird of a particular species might live.

Aristotle's third idea, that too much sex or reproduction results in early death, made such a deep impression that it has remained the focus of attention for centuries. Here is Ray writing in the *Ornithology*:

> The Cock being a most salacious bird doth suddenly grow old, and become less fit for generation. For his spirits being spent, and the radical moisture, as they call it, consumed by the immoderate use of Venery, his body must necessarily wax dry, and his heat of lust be extinguished.[14]

In fact, the natural lifespan of a cockerel was poorly known, for as Ray writes:

> How long these birds would live, were they let alone, I cannot certainly determine, though Aldrovandi limits their age to ten years. For they being kept only for profit, and within a few years ... becoming unfit for generation, who is there that without all hope of gain will keep them only to make experiment how long they will live? But that they are

in their kind short-lived we may rightly infer from their salaciousness and intemperate lust, which infeebles the body, wastes the spirits, and hastens the end.[15]

The debilitating nature of sex was most strikingly confirmed by the extraordinary longevity of castrated individuals, as Ray's colleague Thomas Browne (somewhat cryptically) noted:

> ... yet do we sensibly observe an impotency or total privation thereof, prolongeth life: and they live longest in every kind that exercise it not at all. And this is true not only in eunuch by nature, but spadoes* by art: for castrated animals in every species are longer lived than they which retain their virilities.[16]

What he is saying here is that eunuchs, both natural and man-made, live longer than intact individuals. He was right: the lifespan of human castrates is typically ten or twelve years longer than that of other men.[17]

Aldrovandi also commented that female lifespan could suffer from too much reproduction, as:

> Hens also, sith [sic] they do for the greatest part of the year daily lay eggs, cannot long suffice for so many births, but for the most part after three or four years become barren and effete. For when they have spent all the seed-eggs which from the beginning were in their bodies, they must needs cease to lay, there being no new ones generated within.[18]

This statement actually contains two ideas. The first is that laying too many eggs reduces a hen's lifespan; the second is based on the (correct) notion that females start life with a fixed quota of ova (what Aldrovandi calls seed-eggs) – something that is apparent from the nature

---

* From spayed. The first use of the word in the English language appears in Turberville's *Noble Arte of Venerie or Hunting*, 1576, where there is a chapter on how to spay a bitch to stop her coming into season.

of birds' ovaries with their grape-like bunches of tiny ova: once these have been used up, no new ones can be created.

With the recognition that females as well as males might bear the cost of too much reproduction, the issue switched from the act of copulation itself (mainly with respect to males) to the broader issue of reproduction in general. But it was not until the 1970s that researchers began to think about the link between reproduction and longevity from an evolutionary perspective.[19]

The main difficulty the early ornithologists had in accounting for the different lifespans in different species was a lack of reliable information. There were two problems – almost all the available information came from captive birds, and everyone was fixated on *maximum* lifespan, not recognising that this might provide a distorted view of nature. Ray and Willughby considered birds to be the longest-lived of all warm-blooded animals (relative to their body size), which is true, but some of their longevity estimates were far too high. They reported, for example, that swans might live three hundred years, and they referred to a friend's sound and lusty goose which was 'fourscore years of age'. This seems to be a rare example of sloppy thinking by Ray and Willughby and David Lack much later took them to task for their gullibility, although of course with the benefit of hindsight it was easy to criticise.

A moment of careful reflection would have shown John Ray how unlikely it was that a wild goose might live for eighty years, and the necessary calculations had already been done. In 1646 Thomas Browne, with whose writings Ray should have been familiar, worked out that if a male and female produce a certain number of offspring and their offspring in turn produce a similar number, then within a few generations their descendants would number millions.[20] In other words, for a species that allegedly lived for eighty years and produced a brood of four or five chicks each year, as geese do, their numbers would soon be astronomical. The fact that there weren't this many geese should have alerted Ray to the possibility that something – most probably his estimate or the species' lifespan – was amiss.

Overleaf: Studies of the European robin in the early 1900s, using individually marked wild birds, provided the first reliable estimates of the average lifespan for any species. This painting from 1620–21 is by Vincenzo Leonardi.

Browne used his calculations to develop his ideas about the so-called 'balance of nature', or why animal numbers tended to remain similar over time:

> There are two maine causes of numerosity in any kind of species, that is, frequent and multiparous way of breeding, whereby they fill the world with others, though they exist not long themselves; or a long duration and subsistence, whereby they doe not only replenish the world with a new annumeration of others, but also maintaine the former account of themselves.[21]

In other words there were two broad lifestyles: frequent reproduction and short lifespan, or modest reproduction and extended life. Understanding the balance of nature and what kept the numbers of a particular species from declining to extinction or expanding inexorably was difficult. This is hardly surprising, for with information only on maximum lifespan, Browne, Ray and many others were seriously misled. To make progress, a different kind of thinking was required. The crucial concept was the idea of *average* lifespan but obtaining such information for wild birds required marking and following the fate of a reasonable number of individuals, rather than just one or two.

Some seventeenth-century pioneers like Pernau and Frisch had marked birds by removing toes or attaching coloured threads to their legs but such methods were hardly appropriate for identifying large numbers of individuals. The breakthrough came only around the turn of the twentieth century with the invention of lightweight, individually numbered metal bands. A further refinement, introduced around 1930, was the use of plastic colour rings, employed in different colour combinations that allowed field ornithologists to identify their birds through binoculars without having to catch them. By marking birds as nestlings and following them until they were later found dead, or disappeared, researchers were finally in a position to estimate the average lifespan of particular species.

The first important results about average lifespan came from an amateur ornithologist, James Burkitt, a British engineer who didn't start looking at birds until he was in his thirties and didn't begin his remarkably original study of individually marked robins until he was fifty. Following his birds over successive years during the early 1900s, Burkitt calculated the average lifespan of the robin to be two years and ten months – considerably less than the oldest bird in his study, which was eleven years old when it died.[22]

By following individually recognisable birds in the field Burkitt's studies revealed new secrets about the lives of birds, but he was ahead of his time and his extraordinary results were ignored. It would be a further ten years before Margaret Morse Nice began her now famous study of individually marked song sparrows in Ohio and another decade before David Lack began his own investigation of marked robins in the 1940s and only then were Burkitt's extraordinarily pioneering efforts acknowledged.[23]

Lack's fascination with the European robin marked the beginning of an ambitious, lifelong quest to understand the ecology of bird populations. He is the hero of this chapter, and indeed of modern ornithology, for using natural-selection thinking to successfully understand the life histories of birds.

Educated at Gresham's School in Norfolk and at Magdalene College, Cambridge, between 1929 and 1933, Lack's first employment was as a teacher at Dartington Hall in Devon, where in the liberal environment encouraged by the school he studied the local robins. Lack followed individually marked birds, noting when they disappeared or died so that he could establish their average lifespan.[24] He published his findings in *The Life of the Robin*, a beautifully written book, peppered with wonderful historical anecdotes. Lack's scholarship is apparent on every page, and the book became a classic. Among much else, Lack worked out that he could calculate the average length of life of his colour-ringed robins simply from the proportion of birds that survived from one breeding season to the next. Using simple arithmetic he showed, for example, that if 60 per cent of birds survived between

The life history of the song sparrow was studied by Margaret Morse Nice
in the 1930s. Such was the attitude among professional (i.e. museum) ornithologists
to field studies at that time, her work was first published in Germany rather than
her native United States. (From Audubon's Birds of America, 1827–38)

years the average lifespan must be two years, and if 98 per cent survive the average lifespan would be almost fifty years.[25]

While writing the chapter on survival in his robin book, Lack had a clever idea. He realised that as well as using observations of colour-ringed individuals from detailed field studies like his own, the reports of birds (of whatever species) ringed only with a metal ring and found dead by members of the public could also provide an estimate of a species' average longevity. At last, ornithologists were able to estimate the survival and longevity of a range of different bird species using information already collected by national ringing schemes.

It was ingenious insights like this that eventually made David Lack among the most celebrated of twentieth-century ornithologists. In 1938 he took a year off from teaching to visit the Galapagos Islands and study the ground finches that had so fascinated Darwin a century before. The result was another excellent book on what he called 'Darwin's finches'.[26] During the Second World War, Lack was involved in radar research, which gave him a unique opportunity to use the secret, high-tech equipment then only available to the armed forces, to study bird migration.

After the war Lack was appointed director of the Edward Grey Institute based in the Zoology Department at Oxford. The sequence of events that led to this, and in due course brought about the remarkable modernisation of ornithology, is described by Bill Thorpe:

In 1945 the great change came. In 1927–28 the Oxford Ornithological Society, activated by E. M. Nicholson and others, had founded the Oxford Bird Census, which the following year produced an excellent census of heronries. The obvious merits of this led Max Nicholson and B. W. Tucker to seek some permanent centre for ornithology which would ensure that this admirable census work was not solely dependent upon the comings and goings of a few devoted and able undergraduates. The next step was to appoint W. B. Alexander as Director of the Bird Census. In 1931 a small government grant was found for the work, and in 1932 the British Trust for Ornithology was born with the express

Geospiza magnirostris

The large-billed ground finch from the Galapagos was one of several species studied
by David Lack in 1939. These birds, referred to as Darwin's finches, continue
to provide a wonderful model for evolutionary studies.
(From a painting by John Gould in Darwin, 1839)

object of supporting the Oxford scheme. But funds were short and only in 1938 was it possible for Oxford University formally to back the scheme as a memorial to their late Chancellor, Lord Grey of Fallodon. The new Edward Grey Institute, as it was called, was run almost single-handed by Wilfred Alexander, who continued the census work and built up the magnificent library which is now known by his name, until 1945 when he finally retired. That year A. C. Hardy (the new Oxford Professor of Zoology), B. W. Tucker and A. Landsborough Thomson put forward an overall plan for the full establishment of an Edward Grey Institute of Field Ornithology, as part of a new Department of Zoological Field Studies, and David Lack was appointed as Director.[27]

Lack's remit was to create a 'national coordinating centre for the field ornithology done by amateurs'. However, recognising that his new position gave him the power and potential to change the face of ornithology, Lack ignored these instructions and instead focused on the ecology of birds with natural selection as his guiding principle. By doing so he succeeded in professionalising ornithology.[28]

Instead of applauding Lack's success, however, the original architects of the EGI were dismayed by what they saw as his betrayal. When the British Ornithologists' Union celebrated its centenary in 1959 and published a series of essays in *The Ibis* recounting the recent history of British ornithology, Lack, deliberately or otherwise, gave the impression that mainstream biology had been his goal from the start, and in so doing dismissed at a stroke the efforts of amateur ornithologists. Max Nicholson and James Fisher were appalled and Fisher accused Lack of 'assuming that things that the Institute [EGI] is interested in today are the only things of value'. Part of Fisher's and Nicholson's disappointment was the way that Lack's professional ornithology was 'hyper-critical' of the work of amateurs, which, together with increasing use of statistics in journals like *The Ibis*, effectively excluded amateurs from modern ornithology. To add insult to injury, what Nicholson and Fisher perceived to be Lack's rewriting of EGI history was echoed by other contributors to the *Ibis* centenary volume

– Julian Huxley, Niko Tinbergen, Reg Moreau and Bill Thorpe – who were all of a similar mind to Lack.[29]

Lack revolutionised ornithology, pushing aside the traditional, descriptive studies of bird geography and taxonomy that had dominated the subject for so long, to make way for field studies – the ecology and behaviour of birds. All this happened about twenty-five years after Erwin Stresemann had engineered a similar change in Germany. Stresemann's ideas had little impact in Britain, partly no doubt because of the war and because he usually wrote in German, but also because there was no one in Britain at that time with sufficient vision or intellectual clout to oust the old guard. Lack was eventually that man. The old guard may have been disappointed but they couldn't deny Lack's achievements, and, to be fair, the amateur ornithologists were not completely abandoned: the British Trust for Ornithology (BTO) and the Royal Society for the Protection of Birds (RSPB) quickly assumed a role in engaging them in bird protection and the study of avian ecology.

Soon after becoming the Institute's director in 1945, Lack travelled to the Netherlands to meet the ornithologist Huijbert Kluijver who, since the early 1900s, had been studying a population of great tits breeding in nest boxes. On seeing Kluijver's set-up Lack recognised immediately that this was exactly what he needed to answer his own questions about the population ecology of birds, and on returning to Oxford established his own nest-box population of great tits in Wytham Woods on the outskirts of the city. The study that Lack initiated continues to this day, as does Kluijver's.[30]

One of the first major results of the Oxford great-tit study was the discovery that up to 80 per cent of all young birds die in their first year of life. For many ornithologists and non-ornithologists this was unbelievable (would a benign God be so wasteful?). Lack was heavily criticised: if this many birds were dying, his critics said, we would be knee-deep in tiny corpses. They were wrong and Lack was right and the reason we do not see huge numbers of dead birds is, of course, because they are quickly removed by scavengers or eaten by predators.

David Lack started his ongoing study of the great tit in Wytham Woods near Oxford in the 1940s. (From Selby, 1825–41)

The population studies initiated by Lack bore rich fruit, and within a relatively short time estimates of the survival rates of a great diversity of birds became available. Among other things these studies showed that one important factor affecting how long birds live was their evolutionary history. That is, certain families of birds appear to be predisposed to longer or shorter lives. As we have seen, albatrosses and petrels (also known as tube noses – the Procellariiformes) are among the longest-lived of all birds, even though they vary in size over several orders of magnitude, from tiny storm-petrels, weighing just 35 g, to the wandering albatross that weighs almost three hundred times as much, at 9 kg.

Another observation to emerge from these accumulating studies, and one Lack was quick to notice, was that birds show an unusual pattern of mortality in relation to their age. Instead of the youngest individuals having the highest death rate as occurs in many fish and invertebrates, or an increased likelihood of dying in old age, as in many mammals, for a bird the likelihood of dying remains fairly similar throughout its life. This means, as Aristotle observed, that wild birds rarely show any signs of senescence or age-related reduction in overall performance, and he was right – with the exception of a few isolated examples of exceptionally geriatric *captive* birds, like both the goldfinch mentioned by Gessner and Buffon's nightingale.

As ornithology expanded and researchers started to obtain information about birds in the tropics and other parts of the world, the survival patterns they observed became more complex and more interesting.

A particularly surprising result was that some small tropical passerines had annual survival rates of over 80 per cent and therefore lived just as long as temperate seabirds. And many tropical passerines, like temperate-zone seabirds, also tend to produce rather small broods. This is part of a major geographical pattern and the trend is apparent even within a single species. House sparrows, for example, typically lay a clutch of five eggs in Canada but only two in Central America. Aristotle's link between survival and the amount of energy directed into reproduction – the number of eggs laid, the number of chicks reared – was starting to seem more and more plausible.

In February 1676 John Ray sent his friend Martin Lister a copy of the recently published Latin version of his encyclopaedia. Thanking Ray for the gift, Lister added a few bird notes to his letter, including the fascinating case of a swallow that laid nineteen eggs, instead of the usual clutch of four of five. As each egg was laid Lister removed it, and each day for nineteen days the female laid a replacement egg. Recognising the scientific significance of Lister's remarkable experiment, Ray added this information to the English version of the *Ornithology* that was almost ready for the printers, and later in *The Wisdom of God* expanded on the broader implications of Lister's experiment:

> Birds have not an exact power of numbering, yet have they of distinguishing many from few, and knowing when they come near to a certain number: and that is, that when they have laid such a number of eggs, as they can conveniently cover and hatch.[31]

The question was this: why do different species typically produce clutches of a particular size? Why do albatrosses, razorbills and most other seabirds lay only a single egg in a clutch, while the zebra finch typically lays five eggs and the blue tit ten or twelve? One early idea was that birds laid as many eggs as they were physically capable of, but the fact that they can sometimes lay very large numbers of eggs, as did Lister's swallows, clearly precludes this. Another possibility – and one mentioned by Ray – was that birds laid as many eggs as they could cover and incubate efficiently. Later experiments in which extra eggs were added to clutches showed that this explanation was unlikely since birds could easily incubate more eggs than they usually produced.[32]

Lack's own hypothesis was, like his other ideas, firmly grounded in Darwinian thinking: individuals laid the number of eggs per clutch that resulted in the maximum number of descendants in their lifetime. Ray had said much the same in *The Wisdom of God*: the 'best' clutch

A European swallow that laid nineteen eggs in as many days, as reported to John Ray
by his friend Martin Lister, confirmed that normal clutch size is not determined by
how many eggs a bird is capable of laying. (From Nozeman, 1770–1829)

size was the one most commonly found in nature, and this gave Lack the idea for testing his hypothesis.

What he needed was a bird whose clutch size and fledgling survival were both easily measured. A Swiss study of starlings provided the perfect opportunity. For many years the ornithologist Alfred Schifferli had checked his nest boxes, routinely recording clutch size and ringing all the young starlings before they fledged. Lack knew that, by looking at the subsequent recoveries of these ringed birds and by using the method he had earlier devised, he could estimate their survival. If his clutch-size hypothesis was correct then the proportion of surviving young should be greatest among those starlings laying an average-sized clutch, and less among those laying either fewer or more eggs.

Lack travelled to Sempach in Switzerland, where he and Schifferli sat down to analyse the data. There were no calculators, no computers, and it required a great deal of pencil-and-paper arithmetic to extract the necessary estimates. It was worth it. As the results emerged, the starling data confirmed Lack's hypothesis: starlings laying the commonest clutch size produced the greatest number of surviving young: it was natural selection in action.[33]

Knowing how long different bird species lived on average, or the proportion that survived from one year to the next, was fascinating in its own right, but Lack saw this information as a crucial part of a bigger picture, essential for understanding how bird numbers are maintained. If a population is to remain stable, the death rate must be balanced by the birth rate. Lack took it for granted that bird numbers remain more or less stable, based on the only information then available – the numbers of grey herons, whose population in England and Wales had been counted each year since 1928. The term 'stable' implies exactly the same numbers of birds from one year to the next, but this isn't quite what Lack meant. Rather, what stability means, and as the heron counts showed, was that there was an upper limit, or ceiling, above which numbers did not increase. In fact, heron numbers tended to remain similar unless there was a hard winter, which caused numbers to crash. There were several hard winters during the course of the survey, but that of 1946–7 was one of harshest of the twentieth century, causing a massive decline in heron numbers. Lack noticed, however, that within two or three years their numbers bounced back, indicating that the population clearly had huge potential to expand; but once it reached the level it had been at prior to the hard winter, the increase stopped; the balance was restored.

Belief in 'the balance of nature' is very old indeed. Its origins lie in the Greek idea that nature was both constant and harmonious and that the balance encompassed both the constancy of species and the stability of numbers. Aristotle was no great supporter of the idea, but many of his accounts were consistent with it, including his observation that birds of prey had smaller broods than other types of birds, to maintain the balance of numbers, since, with too many raptors, prey species might become extinct.

Over the ensuing centuries the apparent continuity of nature seemed to confirm the notion that nature was indeed 'balanced' and that some force or process kept the numbers of birds, within a given species, relatively constant. Identifying that process was Lack's goal.

He started by thinking about how heron numbers could recover so quickly after a hard winter and came up with the following scenario. With lakes and ponds covered with ice, the amount of food available for herons is reduced and many die of starvation. Effectively the cupboard is full but the door is locked. Once the ice melts in the spring, the cupboard is unlocked and food becomes available again. But – and this is the crucial point – since the amount of food has *not* been reduced by the winter conditions, the amount now available to each of the surviving herons is actually greater than before. As a result the birds are able to rear more offspring, and the population increases accordingly. However, once the number of herons reaches a level where they again begin to impinge on the food supply – the point at which their birth and death rates balance each other – the numbers of herons stabilise.

The heron population analysis was Lack's model for how all bird populations worked. Food was the key and numbers were regulated in what he referred to as a *density dependent* manner: the more birds there are, the less food there is per individual, and, as a result, the less well they survive or reproduce. While Lack recognised that predators or disease also had the potential to regulate bird numbers, he could not see much evidence for this, and so settled on food as the single most important factor affecting bird populations.[34]

Lack's idea that bird numbers are regulated through density-dependent processes, as illustrated by the heron and great tit, is logical and overtly Darwinian. But not everyone agreed and in particular Vero Wynne-Edwards at the University of Aberdeen saw things differently. Rather than bird numbers being regulated by the environment, via the amount of food available as Lack thought, Wynne-Edwards believed that animals regulated their own numbers through their behaviour. Social displays, such as the gatherings prior to roosting or the assemblages of birds around their breeding colonies, had evolved, Wynne-Edwards proposed, to enable birds to assess their population density and then to act accordingly, keeping their numbers to a level that the food supply would support. When numbers were relatively high, some individuals – usually those lower in the social hierarchy – produced either a smaller clutch or opted not to reproduce at all, thereby helping to keep numbers in check. In this way, Wynne-Edwards suggested, populations never overate their food supply – no one suffered and numbers remained stable.

Wynne-Edwards' idea has an intuitive appeal, and appears to be consistent with many of the facts: socially subordinate individuals often do lay smaller than average clutches, and in some years do not reproduce at all. But it was not the facts that were the issue, it was their interpretation. In truth, the facts were rather few, and there were certainly no experimental data. Good information on population regulation was scarce and Lack's own studies of great tits had not been going long enough by the 1960s to confirm his ideas unequivocally.

The resulting battle between Lack and Wynne-Edwards was as much about rhetoric as it was about data.[35]

Wynne-Edwards' view of the natural world was both naïvely optimistic and wrong. It was wrong because it was based on a mistaken interpretation of the way natural selection operates. He believed individual animals behaved for the good of the species, hence the notion of subordinate individuals sacrificing themselves by cutting back their own reproduction for the good of others. It was the same with territory: as far as Wynne-Edwards was concerned, territory had evolved to space individuals out across the landscape so that those worthy of reproducing got a fair share of the food – again, for the good of the species as a whole. While such self-sacrifice or altruism is morally commendable, nature does not operate on moral principles.

David Lack spent years trying to convince Wynne-Edwards and other ecologists who had not bothered properly to understand the subtlety of natural selection, that this was not how selection works.

The truth is this and the logic is blindingly simple. Imagine a situation where food is in short supply. Now imagine that there are two types of individuals, one adopting Wynne-Edwards' strategy and declining to reproduce, the other following Lack's tactic and breeding as fast as they can. Who has the best chance of getting their genes into the next generation? Lack's tactic is the only one that works: genes that cause individuals to refrain from reproducing cannot be passed on! Natural selection acts on individuals, not groups or populations.

Wynne-Edwards first aired his group-selection views in public in the mid-1950s and initially they were poorly received. At one point it even looked as though they might simply disappear without trace. But with the publication in 1962 of his 653-page tome *Animal Dispersion in Relation to Social Behaviour*, together with his skill as a public speaker, he managed to get group selection back on the agenda. Exasperated, Lack responded both verbally and in writing. At scientific conferences Lack tried to engage Wynne-Edwards in discussion, but he was evasive and never responded, preferring to avoid open conflict.[36] Lack felt frustrated and retaliated by devoting a lengthy part of his next

An annual census of grey herons in England and Wales begun in 1928 provided David Lack with a model for how bird populations are regulated. (From Selby, 1825–41)

book, *Population Studies of Birds*, to explaining why Wynne-Edwards was wrong and why only an individually based selection process was valid, not just in terms of population regulation but in biology as a whole. Lack eventually won, but it took a further decade before group selection was widely realised to be a fallacy.[37]

These two men who spent several years at intellectual loggerheads could not have been more different. Lack was cool, dispassionate and intellectually aggressive. Wynne-Edwards, on the other hand, was quiet and non-confrontational with a warm, outgoing personality. After scientific meetings at which Wynne-Edwards had presented his ideas, Lack came back to Oxford cursing him. But later, when he heard that Lack was planning to take his two sons botanising in the Cairngorm mountains in Scotland, Wynne-Edwards (who was an expert on alpine flora) offered to show them some of the specialities. The trip was a great success and, won over by Wynne-Edwards' charm, Lack returned to Oxford singing his adversary's praises.[38]

We began this chapter by asking why different bird species should have different lifespans. In trying to resolve questions like this Lack was doing something remarkably similar to what Ray had tried to do almost three centuries earlier: drawing on everything he knew about birds to make sense of their lives. Both were fascinated by birds' eggs; their chicks' development and state of development at hatching (blind or precocial); their behaviour (social or solitary); their mating systems (polygamous or monogamous); and their longevity: how all these life-history characteristics fitted together into a single entity that allowed a species to exist in a particular area. Both men sought broad patterns to generate greater understanding.

Lack's goal was ambitious and he knew that, if he was successful, the overall level of understanding would be so much greater than the individual parts. In a way he was combining a collection of different threads – evolution, behaviour, chick-rearing, sociality and so on – and weaving a new multicoloured tapestry of ornithology. At the same time, the story revealed by this tapestry was so exciting that it created new opportunities, new questions, and new perspectives that would carry the understanding of birds' lives even further. Those ornithologists who had gone before had contributed fragments or sometimes even complete threads of comprehension, but Lack's vision was altogether much broader. His clarity of thought regarding evolution (individual selection) convinced him that there were answers and that all he had to do was combine the various pieces of information for it all to fall into place. Of course, he was clever enough to recognise that some of the pieces might not yet be available, and so he encouraged others to generate new ideas and to go out and test them.

An important question for Lack was why different species invested such different amounts of effort into reproduction: why did clutch size differ so much between species? Lack felt that he had satisfactorily answered this question, but his critics were less convinced, arguing that his results were based entirely on observational studies and that experiments were really needed. The required experiment was actually very simple in concept. If Lack's idea of an optimal clutch size was correct, then adding or removing eggs from a clutch should always result in *reduced* breeding success, because *any* deviation from the average number of eggs will produce a sub-optimal clutch size. Lack himself wasn't keen on experiments, preferring natural observations – like the Sempach starling study – but he recognised that they were necessary to test ideas properly. When researchers got round to adding and subtracting eggs in the late 1950s, they were surprised by their results. Contrary to Lack's idea, some species were surprisingly good at raising one or two additional chicks.[39]

This was a puzzle, at least initially, and was resolved only when one further aspect of the birds' biology was taken into account. In building his hypothesis on clutch size Lack had thought only in terms of the current breeding season rather than the birds' entire lifespan and also that those experiments did not take into account the cost of making eggs to the current brood, let alone to future broods. This was an important oversight, for what Lack had not appreciated was that a trade-off exists between clutch size and lifespan – just as Aristotle suggested and as many bird-breeders had known for centuries: too much reproduction *is* bad for you. As clutch-size experiments were conducted on more and more species, it gradually became clear that many birds typically lay fewer eggs than they can actually rear in any one year, to save some energy for future breeding seasons. Incorporating lifespan provided a new, improved version of Lack's idea – natural selection operates not on success during a single breeding season as Lack originally thought, but on a bird's entire *lifetime reproductive success*.

Put another way, natural selection works by favouring those individuals whose entire suite of life-history attributes maximises their lifetime reproductive output in a particular environment. An easy way to think about this is to use Thomas Browne's two extremes – exemplified here by the zebra finch and the razorbill. At the core of the package of life-history attributes is an engine (the bird's body) and a fixed amount of fuel. For the zebra finch the engine runs fast most of the time, but when conditions are right for breeding, the engine runs even faster, flat out in fact, burning fuel at a prodigious rate – effectively putting everything into producing more zebra finches. Even in immature zebra finches the engine runs fast for they become sexually mature at just seventy days. It is hardly surprising then that these birds soon run out of fuel – they live fast and die young, but – and this is the crucial point – this is the most productive strategy for them in the arid regions of Australia in which they live. No other combination of traits would be as successful.

For the razorbill the engine runs more slowly, in low gear; steady, enduring, conserving energy for a long haul. As in other seabirds, maturation occurs slowly and razorbills do not start breeding until they are six or seven years old. Reproduction is carefully paced – a maximum of a single chick per breeding season. Energy conservation is the name of the game. For the razorbills around the wave-washed shores of the North Atlantic, this combination of cautious life-history traits is the most productive. Food is hard to find out at sea so rearing more than one chick per year would be disproportionately demanding. Instead, the best strategy is prudence, retaining some energy for future years.[40]

Thanks to Lack and his many students we now have a clear idea of what makes birds what they are and how all those various features of their lives – their eggs, chicks, songs, plumage, territories and migrations – that so intrigued John Ray fit together into a marvellous web of adaptation.

Does it matter that we understand the lives of birds and whether a particular population of birds is increasing, decreasing or stable? For David Lack and many others, the interest was mainly academic – understanding what maintained the balance of nature was a long-standing intellectual challenge. Answering the puzzle, as we have seen, took centuries of thought and field study. Although John Ray's physico-theology was an improvement on what had gone before and provided an intellectual framework for thinking about the natural world, it failed to account for the balance of nature. The existence of fossil organisms long since extinct implied that God, for all His wisdom, was unable to maintain a perfect balance of nature, leaving Ray disconcerted. Darwin's discovery of natural selection provided a better framework for thinking about these issues, and neatly accounted for Ray's frustrating fossils, but did not immediately account for the balance of nature. A century later Wynne-Edwards thought that he had solved the puzzle, but misunderstood the way natural selection operates – assuming it did so for the good of the species. Lack's individual-selection model, harsh as it may seem, was more logical and provided a much more cohesive account of the facts. Individual-selection thinking also allowed researchers to draw together what had previously seemed a disconnected set of observations. The idea that natural selection operating at the level of the individual optimised suites of life-history characters, including body size, rates of embryo development and age at sexual maturity, was a major step forward in our understanding.

Solving the puzzle of how animal populations remain 'in balance' may initially have arisen from intellectual curiosity, but it also had important practical applications for how we might control pest species, manage those we harvest or save endangered ones. Since Lack's day, the number of endangered bird species has risen inexorably due to man's activities, including the destruction of habitats and, more insidiously, climate change. Although the last century has seen a number of birds become extinct, there have also been some spectacularly successful conservation stories. The Seychelles warbler, for example, went from a handful of pairs in the 1970s to several hundred thirty years later through some innovative tactics employed for their protection. In much the same way, the Néné (Hawaiian goose), the Lord Howe Woodhen, the Mauritius pink pigeon, kestrel and parakeet, and the Seychelles magpie robin have all been brought back from the edge of extinction. None of these conservation success stories would have been possible without understanding bird life histories.

Postscript

John Ray had never enjoyed good health, but the last few years of his life were particularly uncomfortable. Even so he remained busy, revising successive editions of *Wisdom*, writing a guide to the classification of plants, and was working hard to complete the final remnant of a project he started with Willughby, an encyclopaedia of insects.

Ray left Middleton Hall in 1675 after Willughby's mother – his benefactor since his friend's death – had died, moving to Coleshill with his young wife, Margaret, who had been a servant in the Middleton household. When they married Margaret was twenty and Ray forty-five. They lived in several different homes, but when Ray's own mother died in 1679 he took his family, which now included four daughters, back to her old house, Dewlands, at Black Notley in his native Essex. Two years earlier, in 1677, Ray had been offered the prestigious position of secretary to the Royal Society but he declined because he felt it would interfere with his writing. By the turn of the century, after a succession of bitterly cold winters, Ray was suffering dreadfully from diarrhoea, tumours and ulcerated legs. He attributed the ulcers to 'invisible insects' but they were more likely the result of inactivity, poor winter heating and old age. Whatever the cause, Ray was lame and in considerable discomfort: 'You would not imagine that

ulcers of which so little account is made should be so painful and vexatious, they giving me very little respite from pain day and night.'[1]

Ray was in a slow but progressive decline, and further distressed by
the death of one of his twin fourteen-year-old daughters from jaundice
in 1697. It now took all morning to bathe and dress the running sores
on his legs, but he soldiered on and, despite the discomfort, continued
to be astonishingly productive almost to the end. He died on 17 January 1705 and was buried in Black Notley church; he had asked that
'his corpse be nailed up that none might see him'.[2]

When I began this book my ornithological friends nominated David
Lack, Ernst Mayr and Erwin Stresemann as the greatest ornithologists
ever. They were indeed monumental figures and they all acknowledged John Ray's fundamental role in the development of scientific
ornithology. Had it been possible for Ray to meet these three ornithological leaders, I cannot help feeling that his strongest affinity would
have been with David Lack. Not only would they have respected each
other, they probably would have liked each other, too. Ray and Lack
shared a passion for birds; they both believed in God, but believed
more in what they could see with their own eyes and what could be
deduced from a logical, scientific method. Both men were interested
in conceptual issues and shared an uncanny ability to see the big questions in ornithology.

It speaks volumes for Ray's intellectual acuity that he was able to
identify these questions over two hundred years before David Lack
was born. Even more remarkable, as far as I can tell Lack never read
The Wisdom of God.[3] Had he done so, he would no doubt have been
overwhelmed by their joint vision.

I am also certain that, were it somehow possible to have arranged
for the two to meet, Lack would have tried gently but firmly to convince Ray of evolution. And Ray, staunch believer though he was, but
a scientist at heart, would have been persuaded and thrilled by the
incontrovertible fact that natural and sexual selection have shaped
the lives, form and function of birds.

Ray's philosophy allowed him to combine a passion for birds and other aspects of natural history with a deep religious conviction. It seems paradoxical, therefore, that his notion of intelligent design should pave the way for Darwin's ideas. For most people, natural selection came as a shock since not only did it contradict Genesis, creation and the fixity of species, it replaced a benign and intelligent designer with the impersonal process of natural selection. Not all churchmen rejected natural selection though; some, like Charles Kingsley, accepted Darwin's ideas and, as Lack said, 'had they not kept their approval to private letters, they might have helped bridge the widening gap between science and religion'.[4] Some adopted a compromise, accepting that the human body had evolved through natural processes while the soul came as a divine gift. Lack's own view was that, while natural selection was all-important, what set man apart from all other forms of life was a moral sense. Lack simply could not imagine that our ethical or moral values could have evolved through natural selection, and accordingly considered them a divine gift.[5] For Erwin Stresemann and Ernst Mayr, both of whom were atheists, the fact that David Lack was a practising Christian was an anachronism, a curious hangover of the English clergyman-naturalist tradition.[6]

We can see a clear thread running forward from Ray's *Wisdom of God*, through Pernau and Zorn on the Continent, Gilbert White in England, to Darwin, then on to David Lack and finally (at least, so far) to Richard Dawkins. There is a certain irony that the culmination of Ray's physico-theology should be Richard Dawkins' *The Selfish Gene*, for Dawkins is an enthusiastic atheist. But *The Selfish Gene* has changed the way people think about biology, just as Ray's *Wisdom of God* did three hundred years before.

Ray's ideas launched the study of modern ornithology. His encyclopaedia with Willughby was the starting point for the classification of birds, and *The Wisdom of God* the beginning of field studies of birds. The last fifty years have seen the blossoming of both areas of study. Most professional ornithologists were still employed in museums in 1950, but the subsequent expansion of higher education combined with the increasing scientific respectability of ornithology generated a massive rise in the numbers of professional scientists studying birds. The change is reflected in the number of scientific articles on birds published each year: 500 around 1900, 14,000 in 1990[7] and, without doubt, many more since then.

My final image – and it is a fantasy – is of John Ray as the guest of honour at a forthcoming International Ornithological Congress. I can see him listening attentively to the talks, engaging in discussion, and at the end of the meeting, as the grand old man of ornithology, being presented with the set of *Birds of the World*.[8] What would he make of this most comprehensive summary of our current ornithological knowledge? Almost certainly he would be struck by the similarity to his own encyclopaedia in terms of design and the state-of-the-art illustrations. But he would also be amazed by how much we now know, and, of course, by how much more there is still to learn about birds.

Notes

PREFACE

1. Birkhead (1993); Schulze-Hagen et al. (1995, 1999).
2. Key references in the history of ornithology: Newton (1896); Gurney (1921); Allen (1951); Stresemann (1951; 1975); Farber (1982); Barrow (1998); Walters (2003); Burkhardt (2005); Haffer (2001, 2007a); Bircham (2007); for the history of birdwatching see Moss (2004) and Wallace (2004).
3. Cole (1930: 197).
4. White (1954).
5. Kitchell & Resnick's (1999) scholarly analysis of the writings of Albert the Great provide a wonderful example of how ideas can be traced back to their origins.

1. FROM FOLKLORE TO FACTS

1. The kingfisher myth has an ancient origin. Giraldus Cambrensis in the thirteenth century says: 'It is remarkable in these little birds that, if they are preserved in a dry place when dead, they never decay ... What is still more wonderful – if, when dead, they are hung up by their beaks in a dry situation, they change their plumage every year, as if they were restored to life, as though the vital spark still survived ...' Shakespeare cited the kingfisher's prognostication, as does Christopher Marlowe in *The Jew of Malta*: 'Into what corner peers my halcyon's bill.' Remarkably, the belief was still alive in the late nineteenth century (Swainson 1886: 104). The pigeon semen story comes from *El libro de la Utilidades de los Animales*, written in 1354.
2. Armstrong (1958).
3. Browne (1646), cited in Sayle (1927).
4. Ray, in Raven (1942: 467). I have tidied up all quotes from Ray by removing inappropriate capitals and modernising the spelling.
5. Stresemann (1975).
6. Hayes (1972).

7. Stresemann (1975: 3).
8. Darwin to William Ogle, 22 February 1882 (Darwin Correspondence Online Database reference 13697: http://darwin.lib.cam.ac.uk/perl/nav?pclass=calent;pkey=13697).
9. Medawar and Medawar (1984).
10. Hansell (1998); see also Kitchell and Resnick (1999).
11. Charmantier et al., in press.
12. Jonston's decision to exclude emblems from his *Historia Naturalis* (1650–3) has been identified by Ashworth (1990; 1996) as a major step in the development of ornithology. Interestingly, a scarcity of emblems also characteristised another ornithological encyclopaedia from this period, *Traitté General des Oyseaux* by Jean-Baptiste Faultrier (1660). Finding this unpublished manuscript was one of the most exciting discoveries I made during my research for this book. Intended as a gift for Nicolas Fouquet, Louis XIV's minister of finance, Faultrier's encyclopaedia disappeared when Fouquet was clapped in prison on what was probably a trumped-up charge of fraud in 1661. Somehow the manuscript ended up in Lord Derby's library at Knowsley Hall, where I rediscovered it in 2004 (Birkhead et al., 2006a). Subsequent analysis by Isabelle Charmantier has revealed Faultrier's manuscript to be every bit as significant as Jonston's encyclopaedia (Charmantier et. al., in press).
13. Ray (1678: preface). The italics are in the original.
14. Ashworth (1990; 1996).
15. Raven (1942). At thirty-three Ray produced a local flora, the *Cambridge Catalogue* (1660), a British flora (1670), and between 1686 and 1704 three massive volumes describing no fewer than 6,000 plant species. Ray's biographer Charles Raven described these immense volumes as awe-inspiring, but in reality their sheer bulk deterred people from using them and, instead, his much more concise *Methodus plantarum emendata et aucta* (1703), with its ingenious classification system, is regarded as the more useful culmination of his botanical work (see also Pavord, 2005).
16. Derham (1713); see also Arber (1943).
17. Gribbin (2002: 207); see also Mahon (2000).
18. Gurney (1921: 163); Allen (1951: 419–22); Grindle (2005).
19. Raven (1942).
20. Ray's encyclopaedia of 1676 was entitled *Ornithologiae libri tres* (Three books of ornithology), covering: (i) birds in general; (ii) land-fowl, and (iii) water-fowl. This Latin version of the encyclopaedia is usually referred to as the *Ornithologia*. The additional material Ray added to the English edition of 1678 was, Raven (1942) says, to increase sales. The additional information came from explorer-naturalists such as Bontius, Clusius, Hernandez, Marcgrave, Nieremberg, Olina and Piso. A facsimile of the English edition was made from the copy in the Birmingham City Library in 1972.
21. Ray (1678).
22. Jardine (1843: 105, 116).
23. Mayr (1982: 256).
24. Ray (1686, cited in Mayr, 1982: 256).
25. Stresemann (1975).
26. Ray (1678: 12).
27. Ray (1678: 110); Marcgrave's account, *Historia naturalis Brasiliae* (1648). Although it seems unlikely, the ability to change a parrot's plumage turns out to be true. This ex-

traordinary process is referred to as 'tapiragem' and according to ethno-zoologist Teixeira (1985, 1992) the Incas were 'tapirating' parrots over 2,500 years ago, changing their feathers from natural green or blue to yellow or red, the colours they favoured for their ceremonial cloaks and head-dresses. Parrots were kept so that their modified feathers could be harvested through regular plucking. Keen to understand the underlying mechanism Teixeira conducted his own experiments only to discover that repeated plucking alone is sufficient to bring about the change in colour, the secretions from a poison-dart frog – mentioned in several accounts – were unnecessary.

28. Stresemann (1975); note that the anatomist Volcher Coiter made a classification of birds based on structure in 1575.
29. Raven (1942).
30. Stresemann (1975).
31. Ray (1678).
32. Turner (1544).
33. Macleod (1954): of about 156 generic names of British birds, most are attributable to Aristotle and Pliny (about 30 names each), followed by Linnaeus (23), Varro (5), Gessner (4), and Oppian, Apuleius, Homer, Aristophanes and Diogenes one each. For most of the rest, their origin is unclear. Of this sample whose origin is known, more than one-third have their roots in ancient Greece or Rome; Gessner devised a few, as did Ray (e.g. shoveller), and Linnaeus concocted the rest.
34. Macleod (1954).
35. Nutton (1985).
36. Ray (1691).
37. Haffer (1992; 2007a).
38. Ray (1691).
39. Ray (1691).
40. Thomas (1983: 19).
41. Mabey (1986).
42. Ray (1691).
43. Haffer (2001); Birkhead (2003).
44. Zorn's two-volume book *Petino-Theologie* was published in 1742, 1743.
45. Darwin, cited in de Beer (1974: 50).
46. Roger (1997: 312). There is also a story that Linnaeus named the European toad *Bufo bufo* after Buffon, but sadly it doesn't appear to be true. According to Professor Gunnar Broberg at the University of Lund, a leading Linnaeus scholar, Linnaeus had already named the toad before the conflict with Buffon started (Staffan Ulfstrand, personal communication). It is also worth noting that, although Buffon did not write all of the *Natural History of Birds* himself (Roger, 1997; Schmitt, 2007), throughout the text I have referred to him as though he did, for convenience.
47. Mullens (1909), cited in Haffer (2001: 38): it is also deeply ironic that the museum men called part of their work 'field work' – by which they meant the killing and collecting of birds, their nests and eggs.
48. Haffer (2001).
49. Haffer (2004).
50. Haffer (2001: 58).

2. SEEING AND NOT BELIEVING

1. Ray (1691: 16).
2. Ray (1691).
3. The enormous numbers of eggs examined by poultry researchers confirmed the twenty-one-day maximum, but also produced one or two exceptional records of fertile eggs thirty days after the last mating (Romanoff 1960: 95).
4. Fabricius, cited in Adlemann (1942).
5. Harvey, cited in Whitteridge (1981).
6. Fabricius, cited in Adlemann (1942).
7. Harvey, cited in Whitteridge (1981: 184). In fact this was described by Sir Kenelm Digby in 1644, who had access to Harvey's manuscript before it was published in 1651.
8. Harvey, cited in Whitteridge (1981).
9. Ray (1678).
10. Harvey, cited in Whitteridge (1981).
11. Harvey, cited in Whitteridge (1981).
12. Ray (1678: 3).
13. Malpighi made this discovery in 1672 (Malpighi, 1673).
14. Raven (1942: 377).
15. Leeuwenhoek (1678).
16. The 'animalcules' in semen were not called spermatozoa until 1827, when von Baer named them.
17. Ray (1693a).
18. Ray (1691: 118).
19. Ray (1693b).
20. Ray (1693b).
21. More (1653).
22. Ray (1691: 166).
23. Blackburn and Evans (1986); Anderson et al. (1987); Dunbrack and Ramsay (1989).
24. More (1653).
25. Wilson (1991).
26. The Hippocratic writings (cited in Needham, 1959).
27. Schierbeek (1955). Although Coiter described the germinal disc, it was not until 1820 that the Czech physiologist Johannes Purkinje identified the fact that it contains the female pronucleus (Nordenskiold, 1929).
28. Fabricius, cited in Adlemann (1942).
29. Harvey, cited in Whitteridge (1981: 100).
30. Harvey, cited in Whitteridge (1981: 86).
31. As he imagined it – Hartsoeker never said he had actually seen a homunculus; it was what he imagined he would be able to see, if only it were true and he had a sufficiently powerful microscope (Hill, 1985).
32. Wolff (1774).
33. Cobb (2006).
34. Cobb (2006).
35. *Science Illustrated* in 1943; *Saturday Evening Post* in 1950; *The New Yorker* in 1953. Their books are Romanoff & Romanoff (1949), and Romanoff (1960).

36. One of the main questions about embryo development is where different cells come from and how they become muscle, heart or brain. The simple answer is stem cells. These cells have the potential to produce either more stem cells or specific tissues. The study of stem cells is currently a major area of research.
37. Cobb (2006).
38. Lillie (1922).
39. The funnel-like portion of the oviduct in which fertilisation occurs is referred to as the infundibulum. Harper (1904).
40. Olsen and Neher (1948).
41. Ivanoff (1924), cited in Romanoff (1960).
42. Walton and Whetman (1933), cited in Romanoff (1960).
43. Van Drimmelen (1946). His was actually a rediscovery because the occurrence of sperm in the infundibulum had been previously reported by the Danish biologist Peter Tauber in 1875, but had gone unnoticed.
44. Bobr et al. (1964). Although Peter Lake was the first to recognise the significance of the sperm storage tubules for prolonged-fertility birds, he later discovered that Giersberg (1922) had previously noticed the tubules during histological studies, but without appreciating what they were (P. E. Lake, personal communication). In addition, Japanese researchers also discovered sperm storage tubules independently (Fuji & Tamura, 1963).
45. Bray et al. (1975); May and Robertson (1980). It was later confirmed that high levels of extra-pair paternity occurred routinely in red-winged blackbirds and many other bird species (Westneat et al., 1987).
46. S. Hatch, personal communication; see also Hatch (1983).
47. It is more complicated than this since copulation would have to take place an hour or so before ovulation to give the sperm time to travel to the infundibulum.

3. PREPARATION FOR LIFE

1. Lorenz (1935).
2. Pliny (1855).
3. Ray (1678). The name eider duck replaced St Cuthbert's duck from about 1655 (Lockwood, 1984).
4. For a definition see Hess and Petrovitch (1977).
5. Spalding (1873, cited in Hess and Petrovitch, 1977).
6. Spalding (1873, cited in Hess and Petrovitch, 1977).
7. Spalding (1873, cited in Hess and Petrovitch, 1977).
8. Gray (1962). In 1873 Spalding was employed by Lord and Lady Amberley as tutor for their elder son (whose younger brother was Bertrand Russell) at their home. Spalding conducted experiments on ducklings and chicks there, with Lady Kate Amberley acting as research assistant. She also assisted him in other ways. Concerned by Spalding's celibacy, she provided Spalding with 'private instruction in human reproductive biology', bedding Spalding on a regular basis, apparently with her husband's blessing (Boakes, 1984; Richards, 1987).
9. Mascall (1581, cited in Hess and Petrovitch, 1977).
10. See, for example, Frisch (1743–63).
11. Buffon (1778). Here, as elsewhere in the text where I refer to Buffon's *Natural History of Birds*, the dates refer to the original French volume in which that information appears.

12. Hess and Petrovitch (1977).
13. Charmingly described in his popular account of animal behaviour, *King Solomon's Ring* (1952). Originally published as *Er redete mit dem Vieh, den Vögeln und den Fischen* in German in 1949.
14. Zann (1996).
15. Moutjoy et al. (1969); Cade and Burnham (2003).
16. Bateson (1978).
17. Bolhuis (2005).
18. Nicolai (1974); Hauber et al. (2001).
19. Ray (1691: 54).
20. Ray (1678: 16).
21. Koyama (1999).
22. Seibt and Wickler (2006); varied tits – Koyama (1999); food hoarding in tits – Clayton and Cristol (1996).
23. Pliny took this from Aristotle's *History of Animals* (Book IX), but Aristotle merely used the term 'intelligence' to describe the swallow's careful nest construction.
24. Darwin (1871: 101).
25. Ray (1678: 16).
26. Ray (1678: 117).
27. Smellie (1790: 458).
28. Gray (1968).
29. Condillac (1885), cited in Stresemann (1975: 316).
30. Smellie (1790: 144).
31. Julien Offray de la Mettrie, cited in Gray (1968).
32. Leroy (1870).
33. Thorpe (1979).
34. Leroy (1870: 93).
35. Leroy (1870: 70).
36. Leroy (1870: 96).
37. Wallace (1871).
38. Stresemann (1975: 319).
39. Darwin (1871: 104); '[A] high degree of intelligence is certainly compatible with complex instincts.'
40. Darwin (1871: 102).
41. For the golden age of field ornithology, see Haffer (2001).
42. Stresemann (1975).
43. Stresemann (1975). The German ornithological association was then the Deutsche Ornithologische Gesellschaft (DOG); since the Second World War, the Deutsche Ornithologen-Gesellschaft (DO-G).
44. Stresemann (1975).
45. Thorpe (1979); Kruuk (2003); Burkhardt (2005).
46. Others, including Amotz Zahavi and Amanda Ridley, have employed similar methods to Lorenz, with Arabian and pied babblers, respectively, with great success.
47. Kruuk (2003: 218).
48. Alcock (2001); Burkhardt (2005).
49. Hunt (1996); Hunt and Gray (2003); Weir et al. (2002); Bluff et al. (2007).

50. Seibt and Wickler (2006) raised fifty-two goldfinches in captivity.
51. Tebbich et al. (2001).
52. E. Mayr, cited in Haffer (2007b).
53. Dingemanse et al. (2002; 2004).
54. Pepperberg (1999).
55. Emery and Clayton (2004); Dally, Emery and Clayton (2006).

4. DISAPPEARING FANTASIES

1. The research station is Vogelwarte Radolfzell, Germany.
2. Ray (1691).
3. Stanley (1651).
4. Aristotle, *History of Animals* (Book VIII).
5. Exodus xvi: 13; Gurney (1921: 9).
6. Aristotle, *History of Animals* (Book VIII).
7. Belon (1555), Aldrovandi (1599–1603), Topsell (1972) and Faultier (1660) all considered the bat a bird.
8. There seems to be good evidence for birds remaining torpid for a few days, especially migrants caught out by bad weather. McAtee (1947) collated numerous reports of swifts and hirundines (swallows and martins) being found in a torpid state and being 'revived' by the application of warmth (see also Lack, 1956). The most remarkable example of torpor in birds involves the North American Common Poorwill, *Phalaenoptilus nuttalli* (a member of the nightjar family), which can remain in torpor continuously for several weeks during periods of cool weather, effectively hibernating (Woods and Brigham, 2004).
9. Cambrensis (1187).
10. Frederick II, cited in Wood and Fyfe (1943).
11. Kitchell and Resnick (1999, volume 2: 1563).
12. Norderhaug (1984).
13. Denounced as the Antichrist, a heretic and a heathen for challenging the medieval church, Frederick II's writings were prohibited and his magnificent manuscript *The Art of Falconry* remained unpublished until 1596 and virtually unknown until 1788, when two German ornithologists, J. G. Schneider and Blasius Merrem, rediscovered it (Stresemann, 1975; Schramm, 2001). Interestingly, Thomas Browne – one of John Ray's correspondents – knew of it (Keynes, 1964, volume 3: 64), but appears not to have informed Ray.
14. White (1954: 117, 147).
15. Kitching and Resnick (1999).
16. Swainson (1886: 51). The image of swallows being drawn from under the ice of a lake or river, published by Olaus Magnus in his *Historia de Gentibus Septentrionalibus* of 1555, must have helped to fix the idea into people's minds. The idea that swallows overwintered under water did not originate with Olaus Magnus, but was part of popular culture (Gunnar Broberg and Staffan Ulfstrand, personal communication).
17. Stresemann (1975: 286).
18. Hevelius (1666).
19. Buffon (1779).
20. Several authors referred to in Buffon (1779).

21. Southwell (1902), cited in Gurney (1921: 200); see also Sayle (1927).
22. Ray (1678: 212).
23. Ray (1691).
24. The student was Johan Leche (Brusewitz, 1979).
25. Daniel Defoe is an unlikely champion of swallow migration, given that he had been educated at the academy for Dissenters run by the Reverend Charles Morton, who believed that swallows wintered on the moon (see Note 32 below). Defoe's comments were based on his experience in October 1722 at Southwold, Suffolk, during his tour of Great Britain: 'I observ'd in the evening an unusual multitude of birds sitting on the leads of the church; curiosity led me to go nearer and see what they were, and I found they were all swallows ... This led me to enquire of a grave gentleman ... what the meaning was of such a prodigious multitude of swallows sitting there. "Sir" says he, "you are a stranger to it; you must understand first that this is the season of the year when the swallows, their food here failing, begin to leave us, and return to the country, where-ever it may be, from whence I suppose they came ... They are waiting to embark ... the weather being too calm ... they are wind-bound." This was more evident to me, when in the morning I found the wind had come about in the night ... and there was not one swallow to be seen, of near a million, which I believe was there the night before.' (Tour, Letter I 83–5 – cited in Garnett, 1969).
26. Mabey (1986); Barrington (1772).
27. Pennant (1793).
28. Mabey (1986).
29. Barrington (1772).
30. Barrington (1772: 276).
31. Barrington (1772: 276).
32. Morton, in his *Harleian Miscellany* (1744, cited in Gurney, 1921: 200, and Garnett, 1969). Garnett (1969) traced these ideas back to Bishops Godwin and Wilkins, both of whom published accounts (serious or otherwise) of the moon in 1638, and Morton's *Essay towards the probable solution of this Question: Whence come the Stork, and the Turtle* [dove] ... in 1703. In 1714 Cotton Mather of New England, USA, thought the moon was too far, so instead proposed that wild pigeons repaired 'to some undiscovered satellite, accompanying the earth at a near distance' (cited in Allard, 1928). Stillingfleet (1762) also assumed that when birds disappeared in autumn they ascended into the heavens, and that the prognostic abilities of birds were a result of their proximity to God.
33. Berthold (2001).
34. Buffon (1779).
35. Hunter (1786).
36. Barrington (1772: 287)
37. Mabey (1986).
38. Pennant (1768).
39. White (1789).
40. White (1789).
41. Foster (1988) and P. G. M. Foster, personal communication.
42. Mabey (1986).
43. Forster (1808).
44. Anon. (1707).

45. Buffon (1771).
46. The English edition of Buffon (translated by William Smellie (1812)) actually says *sunrise* which is a mistake; the original says sunset.
47. Naumann (1797: 196).
48. Berthold (2001). Johann Andreas Naumann has been usually credited as being the first to describe migratory restlessness but he was preceded by Buffon, Zorn and the author of *Traité du Rossignol* (Anon., 1707).
49. Anon. (1707).
50. Gwinner (1968).
51. Berthold (2001).
52. Van Zomeren (2003).
53. Frisch (1743–63); see Buffon (1779) who also refers to a similar experiment with a piece of brass wire fixed round a swallow's leg.
54. Pernau (1702).
55. Toe-cutting: see Stresemann (1975: 336): see also Jenner (1824). The oldest recorded swift was twenty-one years old when captured alive and released.
56. Berthold (2001).
57. Berthold (2001).
58. S. Emlen, personal communication, 2005.
59. Anon. (1707: 24–6).
60. Middendorf (1859, cited in Berthold, 2001). The first star compass test was conducted by Edgar Sauer in the mid-1950s (see Berthold, 2001), Legg (1780).
61. Perdeck (1958), and personal communication to S. van Balen, 2005.
62. Berthold (2001); P. Berthold, personal communication, 2005, 2006.
63. Berthold (2001).

5. ILLUMINATING DISCOVERIES

1. Anon. (1772); Birkhead (2003).
2. Also in MacPherson (1897), Valli da Todi (1602), Olina (1622), Markham (1621) and Ray (1678).
3. Other terms have been used in different countries: in the Netherlands the process is referred to as 'muiten' – the decoy birds being placed in a 'muit' – equivalent to the word 'mew' in English; in Japan the practice is referred to as 'yogai' (Damsté, 1947); and in Italy in the seventeenth century stopping was used to produce decoy quail, a process referred to as 'la Chiusa alle Quaglie' (Valli da Todi, 1601; MacPherson, 1897: 367).
4. Manzini (1575).
5. Grenze (1938).
6. Including Aldrovandi (1600), Xamarro (1604) and Aitinger (1626).
7. Wickede (1786: chapter 25).
8. Réaumur (1750).
9. Runeberg (1874).
10. Seebohm (1888), cited in Schäfer (1907).
11. Schäfer (1907).
12. Allard (1928).
13. This was a letter to Percy Taverner; 52°[F] = –47°C (cited in Ainley, 1993).

14. Emlen (1969).
15. Damsté started his research in 1937 but, disrupted by the war, he submitted his thesis only in 1946 (Damsté, 1947). Birds' gonads increase in size in spring and decrease in winter (unlike most mammals); this is probably a weight-saving adaptation.
16. Konishi et al. (1989); B. Follett, personal communication.
17. B. Follett, personal communication.
18. Pracontal (2001).
19. Stresemann (1975: 357).
20. Charles Édouard Brown-Séquard was an eminent neurobiologist who in later life dabbled in endocrinology and 'organ therapy'. In a sensational report of 1889 he claimed that by injecting himself with the liquid from the testes of guinea pigs and dogs he had been rejuvenated; he was no longer tired, he could run up and down stairs, and project his jet of urine 25 per cent further than before! His 'organ therapy' became a craze, reaching a peak in the 1930s before being discredited.
21. B. Lofts, personal communication; Parkes (1985).
22. B. Follett, personal communication.
23. Thompson (1924) was also aware of these two types of question.
24. Frederick II, in Wood and Fyfe (1943).
25. Krebs and Davies (1997, 4th edition).
26. Baker (1938: 161).
27. Letter from Mayr to Lack, dated 26 August 1941 (in Haffer (1997)); Johnson (2004: 538–40).
28. One of the earliest and most spectacular calendars recorded the appearance of spring migrants in Norfolk between 1736 and 1810 and 1836 to 1874 – 112 years in total – by four generations of the same family (see Newton, 1896: 557).
29. Jenyns was influenced by the statistical pioneer Lambert-Adolphe-Jacques Quételet.
30. Altum (1868).
31. Lack (1966).

6. THE NOVELTY OF FIELD WORK

1. Ridgway (1901).
2. Burkhardt (2005).
3. Selous (in Stresemann, 1975: 342); Burkhardt (2005).
4. Selous (1933: 136).
5. Burkhardt (2005).
6. Mayr (1935); see also Nice (1933).
7. Burkhardt (2005).
8. Howard (1910: 11).
9. Romanes (1885); Burkhardt (2005: 94).
10. Howard (1910: 8).
11. Morgan (1896).
12. Lack (1959).
13. Howard (1907–14).
14. Huxley (1914).
15. Howard (1908).

16. Burkhardt (2005).
17. Jourdain (1921).
18. Nicholson (1927) says that Howard's *Territory* was remaindered, which is unfortunate, since the original edition is now scarcer than the 1948 reprint, which is among the most commonly encountered bird books in second-hand bookshops; it is both common and affordable and well worth getting.
19. Nicholson (1927).
20. Nicholson (1927).
21. Nicholson (1934).
22. Nice (1941); see also Barrow (1998).
23. Lack and Lack (1933: 197).
24. Lack (1973) – this is Lack's obituary, part of which he wrote himself.
25. Thorpe (1974). Lack may have been more than modest; he may also have been disingenuous. I initially wondered whether the almost venomous, ultra-critical thinking in the Lack and Lack paper of 1933 originated from Lack senior, but later, on hearing the younger David Lack described as 'intellectually combative' (C. M. Perrins, personal communication), I changed my mind and decided that the zealous criticism was just as likely to have been his.
26. Lack (1943).
27. Meise (1930), see also Nice (1941: 449).
28. Haffer (1997: 71).
29. Mayr (1935).
30. Cited with permission of E. Mayr's daughter, Susanne Harrison.
31. Noble (1939).
32. Macdonald (2002: 59).
33. Fisher (1941).
34. Moss (2004), who together with Wallace (2004) provides a superb history of birdwatching.
35. Alexander (1936); Ray (1678: 222).
36. Nice (1941: 442).
37. Olina (1622); Solinas (2000); Birkhead (2003).
38. Manzini (1575) describes the catching of adult birds and rearing young nightingales, but makes no mention of territoriality. Valli da Todi (1601).
39. Valli da Todi (1601).
40. Aristotole, cited in Nice (1953) and Lack (1943). As Nice notes, a succession of others used Aristotle's statement, including Pliny, Frederick II, Gessner, Aldrovandi, Ray and Buffon, all with regard to eagles; what is remarkable is how this seems to have been overlooked.
41. Cited in Lack (1943).
42. This example is from Anon. (1707).
43. Cramp et al. (1988).
44. Gessner (1555).
45. Ticehurst (1934).
46. Leguat (1707); Armstrong (1953); Fuller (2000).
47. Pernau (1707; cited in Stresemann, 1947).
48. Pernau (1707; cited in Stresemann, 1947).

49. Thielcke (1988); see also Stresemann (1947; 1951).
50. Anon. (1728: 24).
51. Albin (1737: 67); note, however, that Albin did not write much that was original so this probably came from someone else, possibly Anon. (1728).
52. White (1789).
53. Pennant (1768).
54. Goldsmith (1774); Nice (1941) suggests that Goldsmith may have been the first to use the term 'territory' with regard to birds.
55. MacPherson (1934).
56. Pitman (1924).
57. Pitman (1924). As MacPherson (1934) has commented, the information on territory is in neither Buffon nor Brisson (1760), so it isn't clear where it came from.
58. Mayr (1935); Altum (1868).
59. Altum (1868).
60. The original scheme of territories was devised by Mayr (1935); Noble (1939); Nice (1941). Later, following a contentious meeting on the topic, Robert Hinde reproduced a somewhat tidier version of Nice's classification (Hinde, 1956).
61. Tinbergen (1939).
62. Lord Tavistock, cited in Nice (1941).
63. Nice (1941).
64. Moffat (1903).
65. Howard (1920); Nicholson (1927).
66. Lack (1943).
67. Wynne-Edwards (1962).
68. For a detailed, lucid account of the issues see Dawkins' *The Selfish Gene* (1976).
69. Lowe (1941).
70. Haffer (1997: 499).
71. Lack (1959).

7. CHORISTERS OF THE GROVES

1. Information on Jürgen Nicolai comes from my e-mail conversations with him, from obituaries by Barlein (2006) and Würdinger (2007), with information on whistling bullfinches from Lichau (1988).
2. Nicolai (1956).
3. Turner (1544); Topsell (1972). I also wonder whether the 'sparrow' of Catullus might have been a bullfinch. Gaius Valerius Catullus (c. 84–54 BC) was a Roman poet who wrote extensively of his love for his mistress Clodia/Lesbia, who seems to have died young (possibly poisoned). His poems were rediscovered in the fifteenth century; his most famous poem is entitled 'Lesbia's Sparrow':

> My lover's sparrow is dead.
> My darling's sparrow is dead,
> Whom she loved more than her own eyes.
> He was honey in her hands, as intimate
> With her as a girl with her own mother.
> That sparrow seldom left her lap.

But pranced upon the air around her, here and there,
Forever making music for his mistress alone.
Now he takes the path of shadows,
Whence neither bird nor man returns.

The identity of Lesbia's sparrow has been much debated, since 'sparrow' indicates any small bird and not necessarily a sparrow such as *Passer domesticus*. Catullus may also have intended a double entendre, for the word 'passera' (little sparrow) in Italian slang refers to the vagina. If it is a bird then the key to its identity may be the Latin word he uses to describe its voice: *pipiabat*, which is unusual (Quinn, 1982) and means either to 'chirp' (in which case it probably was a sparrow) or 'peep' (as in chickens immediately prior to hatching – and also similar to a sparrow's call), or possibly 'pipe' (as translated by Mitchie, 1969), in which case a bullfinch seem possible, for no other bird 'pipes'. The other thing that makes this species possible is the extraordinary devotion hand-reared bullfinches show to their human owners; they are unlike any other bird in this respect.

4. Clement et al. (1993); Arnaiz-Villena et al. (2001), Newton (1972).
5. The reason the bullfinch's sperm are odd-looking is that unconditional monogamy has allowed male bullfinches to cut a few corners and, uniquely among birds, to produce unfinished sperm. As long as they don't have to compete with those of another male, why waste valuable resources finishing sperm off and making them look nice if they do the job as they are? (Birkhead et al., 2006b; Birkhead et al., 2007).
6. Nicolai (1956).
7. Güttinger et al. (2002).
8. Pernau (1707, cited in Stresemann, 1947: 46).
9. Barrington (1773).
10. Nicolai, cited in Barlein (2006).
11. Barrington (1773).
12. Buffon (1778); see also Thorpe (1961).
13. Barrington (1773).
14. Buffon (1770).
15. Bill Thorpe's studies of song-learning among birds reared in auditory isolation were inspired by earlier work by Otto Koehler during the 1950s (cited in Thorpe, 1961). Such birds were referred to as Kaspar Hauser birds, after a boy who had been kept in total isolation since birth and was released on to the streets of Nuremberg as a youth in 1828. The genetic basis of song-learning in canaries is described by Mundinger (1995).
16. For more details on learned versus innate song see Stap's (2005) excellent account.
17. Barrington (1773).
18. Hunter (1786).
19. Aldrovandi (1599–1603); Aristotle, *History of Animals*.
20. King-Hele (1999).
21. Suthers (1990); Larsen and Goller (2002).
22. Aldrovandi (1600).
23. Anon. (1890: 193).
24. The link between the shape of the tongue and ability to speak comes from Aristotle (*History of Animals*, Book II, chapter 12), who also says: 'Certain species of birds above all other animals, and next after man, possess the faculty of uttering articulate sounds; and this faculty is chiefly developed in broad-tongued birds ... The Indian bird, the par-

rot, which is said to have a man's tongue, answers to this description; and, by the way, after drinking wine, the parrot becomes more saucy than ever' [Book VIII, chapter 12]. The Persian physician Avicenna, writing in the tenth century, describes an operation to loosen a child's tongue by cutting the membrane (frenulum) beneath it (Avicenna, 1997). This operation is now considered as unnecessary and ineffectual in humans as it is in birds. The final quote ('To slit their tongues ...') in this paragraph is from Albin (1737).

25. Beckers et al. (2004).
26. Barrington (1773).
27. Recounted to me by one of Nottebohm's students, Art Arnold, personal communication.
28. A. Arnold, personal communication.
29. Marler and Slabbekoorn (2004).
30. Nottebohm et al. (1976); see also Marler and Slabbekoorn (2004).
31. Walton (1653).
32. *The Choristers of the Groves* by Neville Wood (1836); *The Sweet Songsters of Great Britain* by Adams (1860); and *Nature's Music*, edited by Peter Marler and Hans Slabbekoorn (2004).
33. Armstrong (1963: 231). It turns out that this uncharacteristically upbeat passage from Belon (1555) was almost certainly inspired by Jacques Peletier (*Vers Lirique*, 1555); see Glardon (1997).
34. Gardiner (1832), after which Marler's and Slabbekoorn's (2004) book was named.
35. Darwin (1871: 870).
36. West and King (1990).
37. Coxe (1815) – a reprint of part of *The Gentleman's Recreation* by Nicholas Cox [sic], first published in 1674, but see Cox (1677:76).
38. Bechstein (1795).
39. Smith and von Schantz (1993).
40. J. R. Krebs, personal communication.
41. Kircher (1650).
42 Bechstein (1795).
43. Thorpe (1961).
44. Clare wrote the poem in 1832 (Bate, 2003).
45. R. A. Hinde, personal communication. See also Marler and Slabbekoorn (2004); Burkhardt (2005) and http://www.zoo.cam.ac.uk/zoostaff/madingley/history.htm.
46. R. A. Hinde, personal communication.
47. Poulsen (1951); Poulsen was well aware of the song-learning discoveries of Pernau, Barrington and Bechstein.
48. Hinde (1952); Marler (1956).
49. Darwin (1871: 563).
50. Anon. (1707: 36).
51. Buffon (1770).
52. Bechstein (1795).
53. Arnault de Nobleville (1751). Although published anonymously, this is generally thought to be by de Nobleville.
54. Newton (1896: 892), quoting information gleaned from bird-trappers; see also Wood (1836), in which he implies that song is for mate attraction.

55. Montagu (1802); Newton (1896). In the same year that Newton's *Dictionary* appeared, Charles Witchell (1896) published his innovative *Evolution of Bird Song*. A few years later Moffat (1903) states the dual function of birdsong.
56. Craig (1908).
57. Kroodsma (1976).
58. Vallet et al. (1998); see also Marler and Slabbekoorn (2004).
59. Krebs (1977); Krebs et al. (1978).

8. A DELICATE BALANCE

1. Agate et al. (2003). Records of half-siders are reviewed in Taber (1964) and Kumerloeve (1987). My lab produced three gynandromorphs out of 7,500 zebra finches over seventeen years.
2. Forbes (1947).
3. It wasn't unusual in the Middle Ages for animals to be put on trial and executed by the Christian church. Pigs and weevils seem to have been particularly susceptible, weevils for destroying vines and pigs for killing small children (Evans, 1906).
4. Raven (1947: 3).
5. The weasel myth probably had its origin in the fact that the mongoose – a similar-looking animal – could successfully combat snakes (Forbes, 1947). The misericords in Worcester Cathedral date from the late fourteenth century.
6. Evans (1906); Forbes (1947).
7. Lapeyronie subsequently described his findings in a paper read at the Academy of Sciences in 1710 (cited in Evans, 1906).
8. Aristotle, *History of Animals*, 559b, and quoted in White (1954).
9. Aristotle, *History of Animals*, Book VIII.
10. Aldrovandi (1600).
11. Aldrovandi, cited in Lind (1963: 49).
12. Aldrovandi, cited in Lind (1963: 411–12).
13. Harvey, cited in Whitteridge (1981); Lind (1963: 70, 101).
14. Welty (1962).
15. Crew (1965: 72).
16. Cited in Owens and Short (1995); see also Forbes (1947), who refers to a record from the fourth century of a peacock apparently turning into a peahen.
17. Several papers by Yarrell are cited in Forbes (1947).
18. Nordenskiold (1929).
19. Crew (1927: 121).
20. Certain secondary sexual traits, like a cockerel's comb and wattles, *are* controlled by testosterone; and in the zebra finch sexual-plumage dimorphism is not controlled by oestrogen (Arnold, 1996).
21. Selous (1927: 257).
22. Bancke and Meesenburg (1958).
23. Hogan-Warburg (cited in Van Rhijn, 1991).
24. Montagu (1813).
25. Van Oordt and Junge (1936).
26. Lank et al. (1999). Lank's ruff colony was founded using eggs imported from Finland in the 1980s.

27. Jukema and Piersma (2006).
28. Bonnet (1783).
29. Olsen (1965); chickens never showed quite the same tendency to parthenogenesis as turkeys, so it took longer for researchers to produce parthenogenetic chickens; the first one was hatched in 1972.
30. Until recently parthenogenesis was unknown in passerine birds (mainly because no one had looked), until we discovered parthenogenetic eggs in captive zebra finches (Schut, Hemming, Birkhead et al., 2008). Remarkably, as I was writing this I heard of a possible parthenogenetic lovebird in the United States.
31. Ray (1691: 69), but this is not original with Ray; as he acknowledges it is from his friend Dr Ralph Cudworth (1678).
32. Durham and Marryat (1908); see also Mayr (1982: 750).
33. Komdeur et al. (1997).
34. See Olsen and Fraps (1950) for domestic fowl; Stephanie Correa van Veen for zebra finch (personal communication).
35. Balthazart and Adkins-Regan (2002).
36. The terminology is confusing: sex-steroid effects which occur early in development and which are permanent are referred to as 'organisational', whereas those that are temporary are referred to as 'activational'.
37. Agate et al. (2003).

9. DARWIN IN DENIAL

1. Hansell (1998: 142).
2. This law dates from the eleventh century: 'Treatise of Ibn Abdun', Seville. Article 141 (Lévi-Provençal, 1947). The contest known as *triganieri* is clearly an ancient one; it continues to be popular in Scotland in Glasgow's East End (Hansell, 1998; Birkhead, 2000). Darwin (1871) knew about thief pigeons and I find it remarkable that he never followed up on this for it was fabulous evidence for sexual selection; the fact that some males were more attractive than others, but that breeders could artificially select for this attractiveness, confirmed that it was heritable. Darwin was in denial twice over, failing to recognise that females could be promiscuous and that thief pigeons provided incontrovertible evidence for sexual selection.
3. Darwin (1871).
4. Pliny, Book X, chapter 52. In fact, like much else, this comes from Aristotle, who was also probably the first to comment on pigeon infidelity: 'As a general rule these birds [pigeons] show ... conjugal fidelity, but occasionally a female will cohabit with other than her mate' (in Aristotle, *History of Animals*, Book IX).
5. Darwin (1871).
6. Aristotle, *Generation of Animals*, 757b2–3; see also Brock (2004). This statement has caused much discussion.
7. Harvey, cited in Whitteridge (1981: 178).
8. Girton (1765); Tegetmeier (1868) says that Girton was largely plagiarised from Moore (1735).
9. Smellie (1790).
10. Darwin (1871).

11. See Birkhead (2000) and Desmond & Moore (1991). John Murray, Darwin's publisher, felt it was inappropriate to use the term 'sex' on the title page of 'The Descent of Man and Selection in Relation to Sex' (Browne, 2002), and when Darwin did have to refer to something explicitly sexual – such as the colourful swellings of female monkeys' 'bottoms' (in fact their genitalia) – he did so in Latin, presumably to minimise any embarrassment. Intriguingly, when Leeuwenhoek wrote to the Royal Society to tell them about his discovery of (his own) spermatozoa (see chapter 2), he also wrote in Latin – the only time he did so.

12. Buffon (1771).

13. Smellie (1790: 278). Smellie, who translated Buffon's *Histoire Naturelle*, is paraphrasing Buffon here.

14. Moore (1735).

15. Pratt (1852); William Charles Linnaeus Martin's book *Our Domestic Fowls and Song Birds* is undated, but was probably published about 1850. *Linnaeus?* What were his parents thinking?

16. Morris (1853).

17. See, for example, Marler (1956).

18. Williams (1966).

19. Trivers (2002: 58).

20. Sexual selection may even extend beyond fertilisation (see Birkhead and Møller, 1998).

21. Ray (1691: 115).

22. Albert the Great, as cited in Kitchell and Resnick (1999).

23. Evans (1906, 128).

24. Albert the Great, as cited in Kitchell and Resnick (1999: 1565); this may be from Thomas Cantimpre.

25. Harvey, cited in Whitteridge (1981: 40).

26. Harvey, cited in Whitteridge (1981). This wasn't an isolated incident. During the 1960s the French, who were breeding ducks for the foie gras industry, noticed a dramatic drop in their birds' fertility and found the cause to be penis-pecking by the females. Forced to copulate on land rather than water the females treated the males' penises as items of food and tried to eat them (J-P. Brillard, personal communication).

27. Wolfson (1954).

28. Hunter (1786).

29. This is from Estienne and Liebault (1574), *Maison Rustique* (1586 edition: chapter 51).

30. V. Amrhein, personal communication.

31. Anon. (1728).

32. For some reason this is not in the English edition of Hervieux (1718), but is in the French edition of 1713.

33. The dunnock was once a reasonably popular cage-bird in Europe, but I have not seen any reference to its large cloacal protuberance in the early bird literature.

34. The description of the protuberance comes from Fatio (1864), and the first reference to its enormous testes is from Naumann and Naumann (1833, volume 6), who seem to have overlooked the species' cloacal protuberance. Since the birds had been shot it is possible that the protuberance had been damaged.

35. Nakamura (1990); Davies et al. (1995).

36. Birkhead et al. (2006b).

37. Harvey, cited in Whitteridge (1981: 35); this was also commented on by Browne (1646), cited in Keynes (1964, volume 3: 365).

38. Ray (1678: dunnock, p. 168; cock, p. 215; house sparrow, p. 249 – 'the testicles are great'). Others also commented on the differences in testes size between species: Edward Jenner (1824) suggested that species in which the male remains 'but a short time paired with the female' have relatively small testes, 'compared with those that live the connubial state much longer'. He also suggested that species that breed more than once in a season tend to have relatively large testes.

39. Short (1997).

40. Harcourt et al. (1995); Birkhead and Møller (1998).

41. Albert the Great, cited in Kitchell and Resnick (1999: 338, 550).

42. Albert the Great, cited in Kitchell and Resnick (1999: 338).

43. Hunter (1786); see also Moore (2005).

44. Buffon (1781).

45. Baillon was a natural history dealer, best remembered now for a bird – Baillon's crake – named after him by the naturalist Louis Vieillot in 1819 (see Mearns and Mearns, 1988).

46. Møller (1991); see also Birkhead and Møller (1992: 31).

47. Lank et al. (2002).

48. Nordenskiold (1929: 389).

49. Wagner (1836).

50. Afzelius and Baccetti (1991); Retzius (1904–21).

51. Montagu (1802; 1813: 476).

52. Selous (1901); Howard (1907–14).

53. Birkhead and Møller (1992).

54. Smellie (1790: 277); as with much else in Smellie, this is paraphrased from Buffon (1770). Ray (1678: 15) is very brief on this, merely pointing out that some birds pair.

55. Smellie (1790: 227).

56. Lack (1968).

57. Birkhead (1998, 2000).

58. Westneat and Stewart (2003).

10. A DEGENERATE LIFE CORRUPTS

1. Aristotle alludes to this with regard to horses in his *History of Animals*; Aldrovandi (1599–1603) cites Aristotle and Alexander Aphrosiensis on this with respect to brood size and longevity; it reappears in Faultrier (1660); see also Egerton (1975). Albert the Great is quoting Pliny: 'the sparrow ... which has an equal degree of salaciousness [to the ring dove] is short-lived in the extreme ... it is said that the male does not live beyond a year' (Kitchell and Resnick, 1999).

2. Gessner (1555).

3. Manzini (1575).

4. Buffon (1778).

5. Manzini (1575). The Spanish had a monopoly on canaries for around two hundred years after their discovery in 1400.

6. Bacon (1638).

7. Ray (1678: 14).
8. Knapp (1829).
9. Ricklefs (2000a).
10. Mentioned in Thomas Browne (1646) cited in Keynes (1964, volume 2: 1282).
11. Also in Pliny, via William Harvey in *Generation* (1651: see Whitteridge, 1981) and also cited in Ray (1678: 14).
12. Ricklefs (2006).
13. Bennett and Owens (2002); P. Bennett, personal communication; see also Ricklefs (2006).
14. Ray (1678: 155). This is actually from Henry More (1653), who in his *Antidote against Aetheism* says of birds: 'too frequent Venery, [is] ... very prejudicial to their dry carcasses'. Sir Thomas Browne (1646) had made a similar comment in his *Pseudodoxia Epidemica*, but as is clear from Note 1 this statement probably has its roots in Aristotle who believed that longevity was determined by an animal's warmth and moisture – as they age they become colder and drier. See also Egerton (1975: 309).
15. Ray (1678: 155).
16. Browne (1646), cited in Keynes (1964: volume 2: 182); Aristotle also refers to castrates living longer in *History of Animals*.
17. Potts and Short (1999).
18. Aldrovandi (1599–1603).
19. Ricklefs (2000b).
20. Thomas Browne (1646), cited in Keynes (1964); Browne in turn obtained his ideas on this topic from Denis Petau.
21. Browne (1646), cited in Egerton (2005).
22. Burkitt (1924–6).
23. Nice (1937); Lack (1943).
24. In the fourth edition of *The Life of the Robin*, Lack states that Burkitt used colour rings to identify his birds and Nice (1937) says the same thing in her song-sparrow papers. This is wrong: Burkitt used unique permutations of aluminium rings to recognise his birds for the simple reason that he was colour-blind. Colour rings became commercially available only in about 1930 and these are what Lack used to mark his robins individually. Prior to 1930 Mrs Nice had made her own celluloid colour rings. Burkitt's estimate of the average lifespan of his robins was two years and ten months, whereas Lack's was fifteen months, but was calculated in a slightly different way, and comprises the average expectation of further life after the period of post-fledging mortality.
25. Newton (1998). Lack provided a simple formula for converting one to the other: longevity (strictly, the further expectation of life) = $(2-m)/2m$, where m = the average annual mortality expressed as a proportion.
26. Lack (1945).
27. Thorpe (1974); other information on Lack is from: http://www.archiveshub.ac.uk/news/0305lack.html.
28. Johnson (2004).
29. Johnson (2004: 549).
30. Perrins (1979).
31. Ray (1691); Raven (1942).
32. Ricklefs (2000a).

33. Newton (1998).
34. Ignoring disease and predators was a mistake. Lack was correct in assuming food to be important, but subsequent research – some of it on his own great-tit population in Wytham Wood – showed that disease and predators also have a significant role in regulating bird populations (Newton, 1998).
35. Borrello (2003).
36. Borrello (2003). Here was Lack's intellectually combative nature again. Rumour has it that they shouted at each other at conferences, but that is not true. Wynne-Edwards always avoided conflict and never engaged with Lack; instead, when questioned he simply said 'it's all in the book'. Wynne-Edwards (1962); Lack (1966).
37. Dawkins (1976).
38. I. Newton, personal communication.
39. Williams (1986).
40. The relationships Aristotle correctly identified between longevity and (i) reproductive output, (ii) time *in utero*, and (iii) body size are subtle, but work as follows. Producing larger numbers of offspring is energetically demanding and takes energy away from the repair and maintenance of the body, resulting in earlier death. With a finite supply of energy a trade-off exists between longevity and reproduction. Time *in utero* obviously reflects the rate at which an embryo develops. The embryos of long-lived species develop slowly because these species burn fuel at a slow and steady pace from the moment of fertilisation as they develop inside the eggs and, after hatching, throughout life. In addition though, since long-lived birds also tend to be larger, they produce larger eggs and the embryos of larger species take longer to hatch. That greater longevity is associated with larger body size reflects the fact that larger birds are also better able to withstand the elements and to avoid being eaten by predators.

POSTSCRIPT

1. Raven (1942: 279).
2. Mandelbrote (2000); Raven (1942).
3. Lack (1957) cites Thomas Browne's *Religio Medici* (1643) in his *Evolutionary Theory and Christian Belief*, and also knew Ray's biographer, Canon Raven. Thorpe's biographical memoir of Lack for the Royal Society makes no mention of Ray (Thorpe, 1974). I also asked Lack's wife, Elizabeth, and son, Peter, and they were not aware that he had read Ray's *Wisdom* (P. Lack and E. Lack, personal communication).
4. Lack (1957).
5. This is currently a topic of great interest; it is considered by some like R. L. Trivers that ethical values may be subject to natural selection.
6. Gillespie (1987); R. W. Burkhardt, personal communication.
7. Coulson, in Walters (2003: 165).
8. Del Hoyo et al. (1992–2008).

For further information about the writing of this book see: http://www.wisdomofbirds.co.uk

Bibliography

Adams, H. G., *The Sweet Songsters of Great Britain*. London: Gall & Inglis, *c.* 1860.

Adlemann, H. B., *The Embryological Treatises of Hieronymus Fabricius of Aquapendente: The Formation of the Egg and the Chick and the Formed Fetus*. Ithaca: Cornell University Press, 1942.

Afzelius, B., and Baccetti, B., 'History of spermatology.' In Baccetti, B., ed. *Comparative Spermatology 20 Years After*. New York: Raven Press, 1991.

Agate, R. J., Grisham W., Wade J., et al. 'Neural, not gonadal, origin of brain sex differences in a gynandromorphic finch', *Proceedings of the National Academy of Sciences*, 2003, 100: 4873–8.

Ainley, M. G., review of Farber, Paul Lawrence, *The Emergence of Ornithology as a Scientific Discipline: 1760–1850*, *The Auk*, 1983, 100: 763–5.

— *Restless Energy: A Biography of William Rowan 1891–1957*. Montreal: Vehicule Press, 1993.

Aitinger, J. C., *Kurtzer Und Einfeltiger bericht Vom Dem Vogelstellen*. Cassel, 1626.

Albin, E., *A Natural History of English Song-birds*. London: Butterworth & Co., 1737.

Alcock, J., *The Triumph of Sociobiology*. Oxford: Oxford University Press, 2001.

Aldrovandi, U., *Ornithologiae hoc est de avibus historiae*. Bologna, 1599–1603.

Alexander, W. B., '"Territory" recorded for Nightingale in seventeenth century', *British Birds*, 1936, 29: 322–6.

Allard, H. A., 'Bird migration from the point of view of light and length of day changes', *American Naturalist*, 1928, 62: 385–408.

Allen, E., 'The History of American Ornithology before Audubon', *Transactions of the American Philosophical Society*, 1951, 41: 385–591.

Altum, B., *Der Vogel und sein Leben*. Münster, 1868: 240.

Anderson, D. J., Stoyan, N. C., Ricklefs, R. E., 'Why are there no viviparous birds? A comment', *American Naturalist*, 1987, 130: 941–7.

Anon., *Instruction pour elever, nourrir, dresser, instruire et penser toutes sortes de petits oyseaux de volière, que l'on tient en cage pour entendre chanter: avec un petit traite pour les maladies des chiens*. Paris, 1674.

— *Traité du Rossignol*. Paris: Claude Prudhomme, 1707.

— *The Bird-Fancier's Recreation: Being Curious Remarks on the Nature of Song-Birds with choice instructions concerning The taking, feeding, breeding and teaching them, and to know the Cock from the Hen*. London: privately published, 1728.

— *Ornithologia Nova*. Birmingham: Warren, 1745.

— *Unterricht von den verschiedenen Arten der Kanarievögel und der Nachtigallen, wie diese beyderley Vögel aufzuziehen und mit Nutzen so zu paaren seien, dass man schöne Zunge von ihnen haben kann*. Frankfurt and Leipzig, 1772.

— no title. *Avicultura*. 1890, 5: 193.

Arber, A., 'A seventeenth-century naturalist: John Ray', *Isis*, 1943, 34: 319–24.

Aristotle, *History of Animals*.

Armstrong, E. A., 'Territory and birds: a concept which originated from a study of an extinct species', *Discovery*, July 1953: 223–4.

— *The Folklore of Birds*. London: Collins, 1958.

— *A Study of Bird Song*. London: Oxford University Press, 1963.

Arnaiz-Villena, A., Guillén, J., Ruiz-del-Valle, V., et al., 'Phylogeography of crossbills, bullfinches, grosbeaks, and rosefinches', *Cellular and Molecular Life Sciences*, 2001, 58: 1159–66.

Arnault de Nobleville, L. D., *Aedologie, ou Traité du Rossignol Franc, ou Chanteur*, Paris: Debure, 1751.

Arnold, A., 'Genetically triggered sexual differentiation of brain and behavior', *Hormones and Behavior*, 1996, 30: 495–505.

Ashworth, W. B. J., 'Natural history and the emblematic world view.' In Lindberg, D. C., and Westman, R. S., eds, *Reappraisals of the Scientific Revolution*. New York: Cambridge University Press, 1990, 303–32.

— 'Emblematic natural history of the Renaissance.' In Jardine, N., Secord, J. A., and Spary, E. C., eds, *Cultures of Natural History*. Cambridge: Cambridge University Press, 1996.

Avicenna. *The Canon of Medicine*. Tehran: Soroush Press, 1997.

Bacon, F., *Historie Naturall and Experimental, of Life and Death Or of the Prolongation of Life*. London: Lee & Mosely, 1638.

Baker, J. R., *Evolution: Essays on Aspects of Evolutionary Biology*. Oxford: Clarendon Press, 1938.

Balthazart, J., and Adkins-Regan, E., 'Sexual differentiation of brain and behaviour in birds'. In Pfaff, D. W., et al., eds, *Hormones, Brain and Behavior*. San Diego: Academic Press, 2002.

Bancke, P., and Meesenburg, H., 'A study of the display of the ruff (*Philomachus pugnax*)', *Dansk ornithologisk Forenings Tidsskrift*, 1958, 52: 118–41.

Barlein, F. Jürgen Nicolai (1925–2006), *Vogelwarte*, 2006, 44: 193–6.

Barrington, D. H., 'An essay on the periodical appearing and disappearing of certain birds, at different times of the year', *Philosophical Transactions*, 1772, lxii: 265–326.

— 'Experiments and Observations on the singing of Birds', *Philosophical Transactions of the Royal Society of London*, 1773, 63: 249–91.

Barrow, M. V. J., *A Passion for Birds*. Princeton: Princeton University Press, 1998.

Bate, J., *John Clare: A Biography*, London: Picador, 2003.

Bateson, P., 'Sexual imprinting and optimal outbreeding', *Nature*, 1978, 273: 659–60.

Bechstein, J., *Handbuch der Jagdwissenschaft ausgearbeitet nach dem*. Nuremberg: Burgdorfischen, 1801–22.

Bechstein, J. M., *Natural History of Cage Birds*. London: Groombridge, 1795.

— *Gruendliche Anweisung alle Arten von Voegeln zu fangen einzustellen … Neue Auflage*, Nuremburg, 1796.

Beckers, G. J., Nelson, B. S., and Suthers, R. A., 'Vocal-tract filtering by lingual articulation in a parrot', *Current Biology*, 2004, 14: 1592–7.

Belon, P., *L'Histoire de la Nature des Oyseaux*. Paris, 1555.

Bennett, P. M., and Owens, I. P. F., *Evolutionary Ecology of Birds*. Oxford: Oxford University Press, 2002.

Berthold, P., *Bird Migration: A General Survey*. Oxford: Oxford University Press, 2001.

Bewick, T., *History of British Birds*. London: Hodgson, 1797–1804.

Bircham, P., *A History of Ornithology*. London: Collins, 2007.

Birkhead, T. R., 'Avian mating systems: the aquatic warbler is unique', *Trends in Ecology and Evolution*, 1993, 8: 390–1.

— 'Sperm competition in birds: mechanisms and function.' In Birkhead, T. R., and Møller, A. P., eds, *Sperm Competition and Sexual Selection*, London: Academic Press, 1998: 579–622.

— *Promiscuity: An Evolutionary History of Sperm Competition and Sexual Conflict*. London: Faber & Faber, 2000.

— *The Red Canary*. London: Weidenfeld & Nicolson, 2003.

—, Butterworth, E., and van Balen, S., 'A recently discovered seventeenth-century French encyclopaedia of ornithology', *Archives of Natural History*, 2006a, 33: 109–34.

— , Giusti, F., Immler, S., and Jamieson, B. G. M., 'Ultrastructure of the unusual spermatozoon of the Eurasian bullfinch (*Pyrrhula pyrrhula*)', *Acta Zoologica*, 2007, 88: 119–28.

—, Immler, S., Pellatt, E. J., and Freckleton, R., 'Unusual sperm morphology in the Eurasian Bullfinch (*Pyrrhula pyrrhula*)', *The Auk*, 2006b, 123: 383–92.

—, and Møller, A. P., *Sperm Competition in Birds: Evolutionary Causes and Consequences*. London: Academic Press, 1992.

—, eds, *Sperm Competition and Sexual Selection*. London: Academic Press, 1998.

Birkner, W., *Jagdbuch den Vogelherd mit Buschhütte*. Schlagnetz und Lockkäfigen. Unpublished manuscript, 1639.

Blackburn, D. G., and Evans, H. E., 'Why are there no viviparous birds?' *American Naturalist*, 1986, 128: 165–90.

Blackwall, J., 'Tables of the various species of periodical birds observed in the neighbourhood of Manchester with a few remarks tending to establish the opinion that periodical birds migrate', *Memoirs of Literary & Philosophical Society of Manchester*, 1822: 125–50.

Bluff, L. A., Weir, A. A. S., Rutz, C., Wimpenny, J. H., and Kacelnik, A., 'Tool-related cognition in New Caledonian crows', *Comparative Cognition and Behavior Reviews*, 2007, 2: 1–25.

Boakes, R., *From Darwin to Behaviourism*. Cambridge: Cambridge University Press, 1984.

Bobr, L. W., Lorenz, F. W., and Ogasawara, F. X., 'Distribution of spermatozoa in the oviduct and fertility in domestic birds. I. Residence sites of spermatozoa in the fowl oviduct', *Journal of Reproduction and Fertility*, 1964, 8: 3 9–47.

Bolhuis, J. J., 'Development of behaviour.' In Bolhuis, J. J., and Giraldeau, L-A., eds, *The Behavior of Animals*. Oxford: Blackwell, 2005, 119–45.

Bonnet, C., *Oeuvres d'histoire naturelle et de philosophie*. Neuchatel: 1783.

Borrello, M. E., 'Synthesis and selection: Wynne-Edwards' challenge to David Lack', *Journal of the History of Biology*, 2003, 36: 531–66.

Bray, O. E., Kennelly, J. J., and Guarino, J. L., 'Fertility of eggs produced on the territories of vasectomized red-winged blackbirds', *Wilson Bulletin*, 1975, 87: 187–95.

Brisson, M. T., *Ornithologie*. Paris, 1760.

Brock, R., 'Aristotle on sperm competition in birds', *Classic Quarterly*, 2004, 54: 277–8.

Browne, J., *Charles Darwin: The Power of Place*. London: Jonathan Cape, 2002.

Browne, T., *Pseudodoxia Epidemica*, 1646.

Brusewitz, G., *Svalans våta grav*. Ljusdal: E. Ericssons Bokhandel, 1979.

Buckland, F. T., Martin, W. C. L., and Kidd, W., *Birds and Bird Life*. London: Leisure Hour Office, Religious Tract Society, 1859.

Buffon, G. L., *Histoire Naturelle des Oiseaux*. Paris: 1770–83.

Burkhardt, R. W., *Patterns of Behavior: Konrad Lorenz, Niko Tinbergen and the Founding of Ethology*. Chicago: University of Chicago Press, 2005.

Burkitt, J. P., 'A study of robins by means of marked birds', *British Birds*, 1924–6, 17: 294–303; 18: 97–103, 250–7; 19: 120–4; 20: 91–101.

Cade, T. J., and Burnham, W., eds, *Return of the Peregrine*. Boise, ID: The Peregrine Fund, 2003.

Cambrensis, Giraldus, *The Topography of Ireland (Topographia Hibernica)*, 1187.

Catesby, M. *Natural History of Carolina*. London: 1731–43.

— 'Of birds of passage', *Philosophical Transactions of the Royal Society of London*, 1747, 44: 435–44.

Charmantier, I., Greengrass, M., and Birkhead, T. R., 'Jean-Baptiste Faultrier's *Traitté general des Oyseaux* (1660). In press, *Archives of Natural History*.

Clayton, N. S., and Cristol, D. A., 'Effects of photoperiod on memory and food storing in captive marsh tits, *Parus palustris*', *Animal Behaviour*, 1996, 52: 715–26.

Clement, P. H. A., Harris, A., and Davies, J., *Finches and Sparrows: An Identification Guide*. Princeton: Princeton University Press, 1993.

Cobb, M., *The Egg and Sperm Race*. London: Free Press, 2006.

Cole, F. J., *Early Theories of Sexual Generation*. Oxford: Clarendon Press, 1930.

Collinson, P., A letter, *Philosophical Transactions of the Royal Society of London*, 1760, 51: 459–64.

Cornish, J., Letter to Daines Barrington and Dr Maty, *Philosophical Transactions of the Royal Society of London*, 1775, 65: 343–52.

— *Observations on the habits of exotic birds; that is, those which visit England in the spring and retire in the autumn, and those which appear in the autumn and disappear in the spring*. Exeter: W. Norton, 1837.

Coues, E., *Field and General Ornithology*. London: Macmillan, 1890.

Cox, N., *The Gentleman's Recreation*. London: 1677.

Coxe, N., *The Fowler*. London: Dixwell, 1815.

Craig, W., 'The voices of pigeons regarded as a means of social control', *American Journal of Sociology*, 1908, 14: 86–100.

Cramp, S., ed., *Handbook of the Birds of Europe, the Middle East and North Africa*, vol. V. Oxford: Oxford University Press, 1988.

Crew, F. A. E., *The Genetics of Sexuality in Animals*. Cambridge: Cambridge University Press, 1927.

— *Sex Determination*. London: Methuen, 1965.

Cudworth, R., *True Intellectual System of the Universe*. 1678.

Dally, J. M., Emery, N., and Clayton, N. S., 'Food-caching western scrub jays keep track of who was watching when', *Science*, 2006, 312: 1662–5.

Damsté, P. H., 'Experimental modification of the sexual cycle of the greenfinch', *Journal of Experimental Biology*, 1947, 24: 20–35.

Darwin, C., *Birds Part 3 No. 2 of the Zoology of the Voyage of HMS Beagle, by Gould, J.* London: Smith, Elder & Co., 1839.

— *The Descent of Man, and Selection in Relation to Sex.* London: John Murray, 1871.

Davies, N. B., Hartley, I. R., Hatchwell, B. J., Desrochers, A., Skeer, J., and Nobel, D., 'The polygynandrous mating system of the alpine accentor *Prunella collaris*, I. Ecological causes and reproductive conflicts', *Animal Behaviour*, 1995, 49: 769–88.

Dawkins, R., *The Selfish Gene.* Oxford: Oxford University Press, 1976.

De Beer, G., *Charles Darwin; Thomas Henry Huxley: Autobiographies.* London: Oxford University Press, 1974.

Defoe, D., *Tour through the Whole Island of Great Britain*, 1724–7.

Del Hoyo, E. A., ed., *Handbook of Birds of the World.* Barcelona: Lynx Edicions, 1992–.

Derham, W., *Physico-Theology.* 1713.

Desmond, A., and Moore, A., *Darwin.* London: Penguin, 1991.

Dingemanse, N. J., Both, C., Drent, P. J., and Tinbergen, J. M., 'Fitness consequences of avian personalities in a fluctuating environment', *Proceedings of the Royal Society London, Series B*, 2004, 271: 847–52.

—, Both, C., Drent, P. J., van Oers, K., and van Noordwijk, A. J., 'Repeatability and heritability of exploratory behaviour in great tits from the wild', *Animal Behaviour*, 2002, 64: 929–37.

Dresser, H. E., *A History of the Birds of Europe.* London: 1871–81.

Dunbrack, R. L., and Ramsay, M. A., 'The evolution of viviparity in amniote vertebrates: egg retention versus egg size reduction', *American Naturalist*, 1989, 133: 138–48.

Durham, F. M., and Marryat, D. C. E., 'Note on the inheritance of sex in canaries', *Reports to the Evolution Committee of the Royal Society*, 1908, 4: 57–60.

Dursy, E., *Der Primitivstreif.* Lahr: Schauenberg, 1866.

Duval, M., *Atlas d'Embryologie.* Paris: Masson, 1889.

Edwards, G., *Natural History of Birds.* London: 1743–51.

Egerton, F. N., 'Aristotle's population biology', *Arethusa*, 1975, 8: 307–30.

— 'A History of the ecological sciences', Part 15: 'The precocious origins of human and animal demography in the 1600s', *Bulletin of the Ecological Society of America*, 2005, January 2005: 32–8.

Elliot, D. G., *A Monograph of the Birds of Paradise.* London: 1873.

Emery, N., and Clayton, N. S., 'The mentality of crows: convergent evolution of intelligence in corvids and apes', *Science*, 2004, 306: 1903–7.

Emlen, S. T., 'Bird migration: influence of physiological state upon celestial orientation', *Science*, 1969, 165: 716–18.

Estienne, C., and Liebault, J., *L'Agriculture et Maison Rustique.* Paris, 1574.

Evans, E. P., *The Criminal Prosecution and Capital Punishment of Animals.* London: Heinemann, 1906.

Farber, P. L., *The Emergence of Ornithology as a Scientific Discipline: 1760–1850.* Dordrecht: Reidel, 1982.

Fatio, V., 'Note sur une particularité de l'appareil reproducteur mâle chez l'Accentor alpinus', *Revue et Magasin de Zoologie pure et appliquée* (2nd series), 1864, 16: 65–7.

Faultrier, J-B., *Traitté general des Oyseaux.* Paris: Knowsely Hall, 1660, 787.

Fisher, J., *Watching Birds*. Harmondsworth: Penguin, 1941.

Forbes, T. R., 'The crowing hen: early observations on spontaneous sex reversal in birds', *Yale Journal of Biology and Medicine*, 1947, 19: 955–70.

Forster, T. I. G., *Observations on the Brumal Retreat of the Swallow*. London: Philips, 1808.

Foster, P. G. M., *Gilbert White and His Records: A Scientific Biography*. London: Helm, 1988.

Frisch, J. L., *Vorstellung der Vögel Deutschlands, und beyläufig auch einiger Fremden mit ihren natürlichen Farben*, 1743–63.

Fujii, S., and Tamura, T., 'Location of sperms in the oviduct of domestic fowl with special reference to storage of sperms in the vaginal gland', *Journal of the Faculty of Fisheries and Animal Husbandry, Hiroshima University*, 1963, 5: 145–63.

Fuller, E., *Extinct Birds*. New York: Abrams, 2000.

Gardiner, W., *The Music of Nature; or, An Attempt to Prove that what is Passionate and Pleasing in the Art of Singing, Speaking and Performing upon Musical Instruments, is Derived from Sounds of the Animated World*. London, 1832.

Garnett, R., 'Defoe and the swallows', *The Times Literary Supplement*, 1969, 162 (13/2/69).

Gerard, J., *The Herball or Generall historie of plantes*. London, 1597.

Gessner, C., *History of Birds*. Frankfurt, 1555.

Giersberg, H., 'Untersuchungen über Physiologie und Histologie des Eileiters der Reptilien und Vögel; nebst einem Beitrag zur Fasergenese', *Zeitschrift f. wissensch. Zoologie*, 1922, 120, 1–97.

Gillespie, N. C., 'Natural history, natural theology, and social order: John Ray and the "Newtonian Ideology"', *Journal of the History of Natural History*, 1987, 20: 1–49.

Girton, D., *A Treatise on Domestic Pigeons*. London, 1765.

Glardon, P., *Pierre Belon du Mans. L'Histoire de la Nature des Oyseaux. Droz, Genève [Travaux d'Humanisme et Renaissance No. 106]*, 1997.

Goldsmith, O., *An History of the Earth and Animated Nature*. London, 1774.

Gough, J., 'Remarks on the summer birds of passage and on migration in general', *Memoirs of Literary & Philosophical Society of Manchester*, 1812, 2: 453–71.

Gould, J., *The Birds of Europe*. London: 1832–7.

Gray, P. H., 'Douglas Alexander Spalding: the first experimental behaviorist', *Journal of General Psychology*, 1962, 67: 299–307.

— 'The early animal behaviourists', *Isis*, 1968, 59: 372–83.

Grenze, v. d. H., 'Die nightigall-Edelkanarien: Karl Ernst Reich – Bremen über jein lebenswert', *Kanaria*, 1938, week 30: 350–2.

Gribbin, J., *Science, A History*. London: BCA, 2002.

Grindle, N., '"No other sign or note than the very order": Francis Willughby, John Ray and importance of collecting pictures', *Journal of the History of Collections*, 2005, 17: 15–22.

Gurney, J. H., *Early Annals of Ornithology*. London: Witherby, 1921.

Güttinger, H. R., Turner, T., Dobmeyer, S., and Nicolai J., 'Melodiewahrnehmung und Wiedergabe beim Gimpel: Untersuchungen an liederpfeifenden und Kanariengesang imitierenden Gimpeln (*Pyrrhula pyrrhula*)', *Journal für Ornithologie*, 2002, 143: 303–18.

Gwinner, E., 'Artspezifische Muster der Zugunruhe bei Laubsängern und ihre mögliche Bedeutung für die Beendigung des Zuges im Winterquartier', *Zeitschrift für Tierpsychologie*, 1968, 25: 843–53.

Haffer, J., 'The history of species concepts and species limits in ornithology', *Bulletin of the British Ornithologists' Club, Centenary Supplement*, 1992, 112A: 107–58.

— '"We must lead the way on new paths": The work and correspondence of Hartert, Strese-mann, Ernst Mayr – international ornithologists', *Ökologie der Vögel*, 1997, 19: 3–980.

— 'Erwin Stresemann (1889–1972) – Life and work of a pioneer of scientific ornithology: a survey.' In Haffer, J., Rutschke, E., and Wunderlich, K., eds, *Erwin Stresemann (1889–1972). Leben und Werk eines Pioniers der wissenschaftlichen Ornithologie. Acta Historica Leopoldina*, 34 (Deutsche Akademie der Naturforscher), Stuttgart: Wissenschaftliche Verlagsgesells-chaft, 2000.

— 'Ornithological research traditions in central Europe during the 19th and 20th centuries', *Journal für Ornithologie*, 2001, 142: 27–93.

— 'Erwin Stresemann (1889–1972) – Life and work of a pioneer in scientific ornithology: A survey', *Acta Historica Leopoldina*, 2004, 34:1–465.

— 'Altmeister der Feld-Ornithologie in Deutschland', *Blatter aus dem Naumann-Museum*, 2006, 25: 1–55.

— 'The development of ornithology in central Europe', *Journal für Ornithologie*, 2007a, 148: S125–S153.

— *Ornithology, Evolution and Philosophy: The Life and Science of Ernst Mayr (1904–2005)*. Berlin and Heidelberg: Springer Verlag, 2007b.

Hansell, J., *The Pigeon in History*. Bath: Millstream, 1998.

Harcourt, A. H., Purvis, A., and Liles, L., 'Sperm competition: mating system, not breeding season, affects testes size of primates', *Functional Ecology*, 1995, 9: 468–76.

Harper, E., 'The fertilization and early development of the pigeon's egg', *American Journal of Anatomy*, 1904, 3: 349–86.

Harvey, P. H., and Pagel, M. D., *The Comparative Method in Evolutionary Biology*, Oxford: Oxford University Press, 1991.

Hatch, S. A., 'Mechanism and ecological significance of sperm storage in the northern fulmar with reference to its occurrence in other birds', *The Auk*, 1983, 100: 593–600.

Hauber, M. E., Russo, S. A., and Sherman, P. W., 'A password for species recognition in a brood-parasitic bird', *Proceedings of the Royal Society of London. Series B: Biological Sciences*, 2001, 268: 1041–8.

Hayes, H. R., *Birds, Beasts and Men*. Dent: London, 1972.

Hayes, W., *A Natural History of British Birds*. London: Hooper, 1771–5.

Hervieux de Chanteloup, J-C., *Nouveau Traité des Serins de Canarie*. Paris: Claude Prud-homme, 1713.

— *A New Treatise of Canary Birds*. London: Bernard Lintot, 1718.

Hess, E. H., and Petrovich, S. B., eds, *Imprinting*, volume 5. Stroudsurg: Dowden, Hutchinson & Ross, 1977.

Hevelius, J., 'Promiscuous enquiries, chiefly about cold, formerly sent and recommended to Monsieur Hevelius; together with his answer return'd to some of them', *Philosophical Trans-actions of the Royal Society of London*, 1666, 19: 350.

Hill, K. A., 'Hartsoeker's homunculus: A corrective note', *Journal of the History of the Behav-ioral Sciences*, 1985, 21: 178–9.

Hinde, R. A., 'The behaviour of the great tit (*Parus major*) and some other related species', *Behaviour, Suppl.*, 1952, 2: 1–201.

— 'The biological significance of territories in birds', *The Ibis*, 1956, 98: 340–69.

Hogan-Warburg, L., 'Social behaviour of the ruff *Philomachus pugnax*', *Ardea*, 1966, 54: 109–229.

Howard, E., *The British Warblers*. London: Porter, 1907–14.

— *Territory in Bird Life*. London: Murray, 1920.

— *An Introduction to the Study of Bird Behaviour*. Cambridge: Cambridge University Press, 1929.

Hunt, G. R., 'Manufacture and use of hook-tools by New Caledonian crows', *Nature*, 1996, 379: 249–51.

— and Gray, R. D., 'Diversification and cumulative evolution in New Caledonian crow tool manufacture', *Proceedings of the Royal Society of London. Series B: Biological Sciences*, 2003, 270: 867–74.

Hunter, J., *Observations on certain parts of the animal economy*, 2nd edn. London, 1786.

Huxley, J. S., 'The courtship habits of the great crested grebe (*Podiceps cristatus*); with an addition to the theory of sexual selection', *Proceedings of the Zoological Society of London*, 1914, 2: 491–562.

Jardine, W., 'Memoir of Francis Willughby', *The Naturalist's Library*, 1843, 36: 17–146.

Jenner, E., 'Some observations on the migration of birds', *Philosophical Transactions of the Royal Society of London*, 1824, 42: 11–44.

Jenyns, L., *Observations in Natural History*. London: Van Voorst, 1846.

Johnson, K., '*The Ibis*: transformations in a twentieth-century British natural history journal', *Journal of the History of Biology*, 2004, 37: 515–55.

Jonston, J. *Historia Naturalis*, Frankfurt, 1650–3 (English translation, 1657).

Jourdain, F. C. R., 'Howard on territory in bird life', *The Ibis*, 1921 (no volume): 322–4.

Jukema, J., and Piersma, T., 'Permanent female mimics in a lekking shorebird', *Biology Letters*, 2006, 2: 161–4.

Keynes, G., *The Works of Sir Thomas Browne*. London: Faber & Faber, 1964.

King-Hele, D., *Erasmus Darwin*. London: de la Mare, 1999.

Kinzelbach, R. K., and Hölzinger, J., *Marcus zum Lamm (1544–1606): Die Vogelbucher aus dem Thesaurus Picturarum*. Stuttgart: Eugen Ulmer, 2001.

Kircher, A., *Musurgia Universalis*. Rome: Corbetti, 1650.

Kitchell, K. F., and Resnick, I. M., *Albertus Magnus on Animals: A Medieval Summa Zoologica*. Baltimore: Johns Hopkins University Press, 1999.

Klein, J. T., *Historiae avium prodromus*. Lübeck: Schmidt, 1750.

Knapp, J. L., *The Journal of a Naturalist*. London: John Murray, 1829.

Komdeur, J., Daan, S., Tinbergen, J., and Mateman, C., 'Extreme adaptive modification in sex ratio of the Seychelles warbler's eggs', *Nature*, 1997, 385: 522–5.

Konishi, M., Emlen, S. T., Ricklefs, R. E., and Wingfield, J. C., 'Contributions of bird studies to biology', *Science*, 1989, 246: 465–72.

Koyama, S., *Tricks using Varied Tits: Its History and Structure* [in Japanese]. Tokyo: Hosei University Press, 1999.

Krebs, J. R., 'The significance of song repertoires: the Beau Geste hypothesis', *Animal Behaviour*, 1977, 25: 428–78.

—, Ashcroft, R., and Webber, M., 'Song repertoires and territory defence in the great tit', *Nature*, 1978, 271: 539–42.

— and Davies, N. B., *Behavioural Ecology: An Evolutionary Approach*, 4th edn. Oxford: Blackwell Scientific Publications, 1997.

Kroodsma, D. E., 'Reproductive development in a female songbird: Differential stimulation by quality of male song', *Science*, 1976, 192: 574–5.

Kruuk, H., *Niko's Nature: A Life of Niko Tinbergen*. Oxford: Oxford University Press, 2003.

Kumerloeve, H., 'Le gynandromorphisme chez les oiseaux. Recapitulation des données connues', *Alauda*, 1987, 55: 1–9.

Lack, D., *The Life of the Robin*. London: Witherby, 1943.

— *Darwin's Finches*. Cambridge: Cambridge University Press, 1945.

— *Swifts in a Tower*. London: Methuen, 1956.

— *Evolutionary Theory and Christian Belief*. London: Methuen, 1957.

— 'Some British pioneers in ornithological research 1859–1939', *The Ibis*, 1959, 101: 71–81.

— *Population Studies of Birds*. Oxford: Clarendon Press, 1966.

— *Ecological Adaptations for Breeding in Birds*. London: Chapman & Hall, 1968.

— 'My life as an amateur ornithologist', *The Ibis*, 1973, 115: 421–31.

— and Lack, L., 'Territory reviewed', *British Birds*, 1933, 27: 179–99.

Lank, D. B., Coupe, M., and Wynne-Edwards, K. E., 'Testosterone-induced male traits in female ruffs (*Philomachus pugnax*): autosomal inheritance and gender differentiation', *Proceedings of the Royal Society of London. Series B: Biological Sciences*, 1999, 266: 2323–30.

—, Smith, C. M., Hanotte, O., Ohtohen, A., Bailey, S., and Burke, T., 'High frequency of polyandry in a lek mating system', *Behavioral Ecology*, 2002, 13: 209–15.

Larsen, O., and Goller, F., 'Direct observation of syringeal muscle function in songbirds and a parrot', *Journal of Experimental Biology*, 2002, 205: 25–35.

Leeuwenhoek, A., 'Observationes D. Anthonii Lewenhoeck, de Natis è semine genitali Animalculis', *Phil. Trans Roy. Soc. London*, 1678, 12: 1040–3.

Legg, J., *A discourse on the emigration of British birds …* London: Fielding and Walker, 1780.

Leguat, F., *Voyages des aventures de François Leguat et de ses compagnons en deux isles désertes des Indes orientales*. Paris, 1707.

Leroy, G. C., *The Intelligence and affectability of animals from a Philosophic Point of View, with a few Letters on Man*. London: Chapman & Hall, 1870.

Levaillant, F., *Histoire Naturelle des Perroquets*. Paris, 1801–5.

Lévi-Provençal, E., *Seville Musulmane au début du XII siècle: le Traité d'Ibn Abdun sur la vie urbaine et les corps de métiers*. Paris: Maisonneuve, 1947.

Lichau, K-H., 'Zur Geschichte der liederpfeifenden Dompfaffe im Vogelsberg', *Die Gefiederte Welt*, 1988, 112: 17–19, 45–7.

Lillie, F. R., 'Charles Otis Whitman', *Journal of Morphology*, 1911, 22: 14–77.

Lind, L. R., ed., *Aldrovandi on Chickens: The Ornithology of Ulisse Aldrovandi (1600)*, volume II, Book XIV. Norman, OK: University of Oklahoma Press, 1963.

Lockwood, W. B., *The Oxford Book of British Bird Names*, Oxford: Oxford University Press, 1984.

Lorenz, K. Z., 'Der Kumpan in der Umwelt des Vogels: der Artgenosse als aus ösendes Moment sozialer Verhaltensweisen', *Journal für Ornithologie*, 1935, 83: 137–15.

— *King Solomon's Ring*. London: Methuen, 1952.

Lowe, P., 'Henry Eliot Howard: an appreciation', *British Birds*, 1941, 34: 195–7.

Mabey, R., *Gilbert White*. London: Hutchinson, 1986.

Macdonald, H., '"What makes you a scientist is the way you look at things": ornithology and the observer 1930–1955', *Studies in History and Philosophy of Science. Part C: Studies in History and Philosophy of Biological and Biomedical Sciences*, 2002, 33: 53–77.

Macleod, R. D., *Key to the Names of British Birds*. London: Pitman, 1954.

MacPherson, A. H., 'Territory in bird life', *British Birds*, 1934, 27: 266.

MacPherson, H. A., *A History of Fowling*. Edinburgh: D. Douglas, 1897.

Mahon, S., 'John Ray (1627–1705) and the act of uniformity 1662', *Notes and Records of the Royal Society of London*, 2000, 54: 153–78.

Malpighi, M., *De formatione pulli in ovo* [*On the formation of the chicken in the egg*], 1673.

Mandelbrote, S., 'John Ray.' In Matthew, H. C. G., and Harrison, B., eds, *Oxford Dictionary of National Biography*, volume 46. Oxford: Oxford University Press, 2004, 178–83.

Manzini, C., *Ammaestramenti per allevare, pascere, and curare gli ucceli*. Brescia: Pietro Maria Marchetti, 1575.

Marcgrave, G., *Historia Naturalis Brasiliae*. 1648.

Markham, G., *Hungers Prevention: or The Whole Art of Fowling by Water and Land*. London: Francis Grove, 1621.

Marler, P., 'Behaviour of the chaffinch *Fringilla coelebs*. *Behaviour*, *Suppl.*, 1956, 5: 1–184.

—, and Slabbekoorn, H. *Nature's Music: The Science of Birdsong*. Amsterdam: Elsevier, 2004.

Martin, W. C. L., *Our Domestic Fowls and Song Birds*. London: Religious Tract Society, n.d.

Mason, E. A., 'Determining sex in breeding birds', *Bird Banding*, 1938, 9: 46–8.

May, R. M., and Robertson, M., 'Just so stories and cautionary tales', *Nature*, 1980, 286: 327–9.

Mayr, E., 'Bernard Altum and the territory theory', *Proceedings of the Linnaean Society of New York*, 1935, 45/46: 24–38.

— *The Growth of Biological Thought*. Cambridge, MA: Belknap Press, 1982.

McAtee, W. L., 'Torpidity in birds', *American Midland Naturalist*, 1947, 38: 191–206.

Mearns, B., and Mearns R., *Biographies for Birdwatchers*. London: Academic Press, 1988.

Medawar, P. B., and Medawar, J. S., *Aristotle to Zoos*. London: Weidenfeld & Nicolson, 1984.

Megenberg, C., v., *Das Buch der Natur*. 1358.

Meise, W., 'Revierbesitz im Vogelleben', *Eine Umschau. – Mitteilungen des Vereins Saechsischer Ornithologen*, 1930, 3: 49–68.

Meyer, H. L., *Coloured illustrations of British Birds, and their eggs*. London: Longman, 1835–50.

Meyer, J. D., *Angeenehmer und nützlicher Zeit-Vertreib mit Betrachtung*. Nuremberg: 1748–56.

Michie, J., *The Poems of Catullus*. London: Rupert Hart-Davis, 1969.

Moffat, C. B., 'The spring rivalry of birds: some views on the limit to multiplication', *Irish Naturalist*, 1903, 12: 152–66.

Montagu, G., *Ornithological Dictionary*. London: White, 1802.

— *Supplement to the Ornithological Dictionary*. Exeter: Woolmer, 1813.

Moore, J., *Columbarium, or the Pigeon House – Being an Introduction to a Natural History of Tame Pigeons*. London: Wilford, 1735.

Moore, W., *The Knife Man*. London: Bantam Press, 2005.

More, H., *An antidote against atheism: or an appeal to the natural faculties of the minds of man, whether there be not a God*. London: Daniel, 1653.

Morgan, C. L., *Habit and Instinct*. London: Arnold, 1896.

Morris, F. O., *A History of British Birds*. London: Groombridge, 1851–7.

Morton, C., *An Essay towards the probable solution of this Question, whence come the stork, and the turtle dove … Samuel Crouch, 1703.

— *The Harleian Miscellany*. 1744.

Moss, S., *A Bird in the Bush: A Social History of Birdwatching*. London: Aurum, 2004.

Mountjoy, P. T., Bos, J. H., Duncan, M. O., and Verplank, R. B. 'Falconry: neglected aspect of the history of psychology', *Journal of the History of the Behavioral Sciences*, 1969, 5: 59–67.

Møller, A. P., 'Sperm competition, sperm depletion, parental care and relative testis size in birds', *American Naturalist*, 1991, 137: 882–906.

Mudie, R., *The British Naturalist*. London: Orr & Smith, 1830.

Mullens, W. H., *Some early British ornithologists and their works*. IX William MacGillivray, W. (1796–1852), and Yarrell, W. (1784–1853), *British Birds*, 1909, 2: 389–99.

Muller, J., Bell, F. J., and Garrod, A. H., *On certain variations in the vocal organs of the Passeres that have hitherto escaped notice*. Oxford: 1878.

Mundinger, P. C., 'Behavior-genetic analysis of canary song: inter-strain differences in sensory learning, and epigenetic rules', *Animal Behaviour*, 1995, 50: 1491–1511.

Nakamura, M., 'Cloacal protuberance and copulatory behaviour of the Alpine accentor (*Prunella collaris*)', *The Auk*, 1990, 107: 284–95.

Naumann, J. A., *Naturgeschicte der Land- und Wasser-Vögel des nordlichen Deurschlands*. Kothen, 1795–1803.

— and Naumann, J. F., *Naturgeschichte der Vögel Deutschlands*. 1820–60.

Naumann, J. F., *Naturgeschichte der Vögel Mitteleuropas*. Vol. IV. Gera-Untermhaus, Kohle. 1905.

Needham, J., *A History of Embryology*. London: Abelard-Schuman, 1959.

Newton, A., *A Dictionary of Birds*. London: A&C Black, 1896.

Newton, I., *Finches*. London: Collins, 1972.

— *Population Limitation in Birds*. San Diego: Academic Press, 1998.

Nice, M. M., 'The theory of territorialism and its development.' In Chapman, F. M., and Palmer, T. S., eds, *Fifty years' progress of American ornithology, 1883–1933*, Lancaster, PA: American Ornithologists' Union, 1933, 89–100.

— 'Studies in the life history of the song sparrow', *Transactions of the Linnaean Society, New York*, 1937, 4: 1–247.

— 'The role of territory in bird life', *American Midland Naturalist*, 1941, 26: 441–87.

— 'The earliest mention of territory', *Condor*, 1953, 55: 316–17.

Nicholson, E. M., *How Birds Live*. London: Williams & Norgate, 1927.

— 'Territory reviewed', *British Birds*, 1934, 27: 234–6.

Nicolai, J., 'Zur Biologie und Ethologie des Gimpels (*Pyrrhula pyrrhula* L.)', *Zeitschrift für Tierpsychologie*, 1956, 13: 93–132.

Nicolai, J., 'Mimicry in parasitic birds', *Scientific American*, 1974, 231: 92–8.

Noble, G. K., 'The role of dominance in the life of birds', *The Auk*, 1939, 56: 263–73.

Nordenskiold, E., *The History of Biology: A Survey*. London: Paul, Trench & Trubner, 1929.

Norderhaug, M., 'The Svalbard Geese: an introductory review of research and conservation', *Norsk Polarinstitutt Skrifter*, 1984, 181: 7–10.

Nottebohm, F., Stokes, T. M., and Leonard, C. M., 'Central control of song in the canary, *Serinus canarius*', *Journal of Comparative Neurology*, 1976, 165: 457–86.

Nozeman, C., *Nederlansche vogelen*. Amsterdam: Sepp, 1770–1829.

Nutton, V., 'Conrad Gesner and the English naturalists', *Medical History*, 1985, 29: 93–7.

Olaus Magnus, *Historia de Gentibus Septentrionalibus*. Rome, 1555.

Olina, G. P., *L'Uccelliera*. Rome: 1622.

Olsen, M. W., 'Twelve-year summary of selection for parthenogenesis in Beltsville small white turkeys', *British Poultry Science*, 1965, 6: 1–6.

— and Fraps, R. M., 'Maturation changes in the hen's ovum', *Journal of Experimental Zoology*, 1950, 144: 475–87.

— and Neher, B. H., 'The site of fertilization in the domestic fowl', *Journal of Experimental Zoology*, 1948, 109: 355–66.

Owen, C., *An Essay Towards a Natural History of Serpents*. London, 1742.

Owens, I. P. F., and Short, R. V., 'Hormonal basis of sexual dimorphism in birds: implications of new theories of sexual selection', *Trends in Ecology & Evolution*, 1995, 10: 44–7.

Parkes, A. S., *Off-beat Biologist: The Autobiography of Alan S. Parkes*. Cambridge: Galton Foundation, 1985.

Pavord, A., *The Naming of Names*. London: Bloomsbury, 2005.

Pennant, T., *British Zoology*, London: Benjamin White, 1768.

— *The Literary Life of the Late Thomas Pennant, Esq., by Himself*, London, 1793.

Pepperberg, I. M., *The Alex Studies*. Harvard: Harvard University Press, 1999.

Perdeck, A. C., 'Two types of orientation in migrating Starlings, *Sturnus vulgaris* L., and chaffinches, *Fringilla coelebs* L., as revealed by displacement experiments', *Ardea*, 1958, 46: 1–37.

Pernau, F. A. v., *Unterricht was mit dem lieblichen Geschöpff, denen Vögeln*. 1702.

Perrins, C. M., *British Tits*. London: Collins, 1979.

Pitman, J. H., *Goldsmith's Animated Nature: A Study of Goldsmith*. Yale: University of Yale, 1924.

Pliny, *Naturalis Historia*, Book X: *The Natural History of Birds*. London: Taylor & Francis, 1855.

Potts, M., and Short, R. V., *Ever Since Adam and Eve*. Cambridge: Cambridge University Press, 1999.

Poulsen, H., 'Inheritance and learning in the song of the chaffinch (*Fringilla coelebs*)', *Behaviour*, 1951, 3: 216–28.

Pracontal, M. de, *L'imposture scientifique en dix leçons*. Paris: Decouverte, 2001.

Pratt, A., *Our Native Songsters*. London: SPCK, 1852.

Quinn, K., *Catullus, The Poems*, London: Macmillan, 1982.

Raven, C. E., *John Ray, Naturalist: His Life and Works*. Cambridge: Cambridge University Press, 1942.

— *English Naturalists from Neckam to Ray*. Cambridge: Cambridge University Press, 1947: 379.

Ray, J., *The Ornithology of Francis Willughby*. London: John Martyn, 1678.

— *The Wisdom of God Manifested in the Works of Creation*. London: Smith, 1691.

— *Synopsis Animalium Quadrupedum*. London: Smith, 1693a.

— *Three Physico-Theological Discourses*. London: Smith, 1693b.

Réaumur, M. de, *The Art of Hatching and Bringing up Domestick Fowles of all kinds, at any time of year, either by means of hot-beds, or that of common fire*. Paris: Royal Academy of Sciences, 1750.

Rem, G., *Emblematica Politica*. 1617.

Rennie, J., *The Faculties of Birds*. London: Knight, 1835.

Retzius, G., *Biologische Untersuchungen, Neue Folge*. Stockholm & Leipzig: 1904–21.

Richards, R. J., *Darwin and the Emergence of Evolutionary Theories of Mind and Behavior*. Chicago: University of Chicago Press, 1987.

Ricklefs, R. E., 'Lack, Skutch, and Moreau: the early development of life-history thinking', *Condor*, 2000a, 102: 3–8.

— 'Intrinsic aging-related mortality in birds', *Journal of Avian Biology*, 2000b, 31: 103–11.

— 'Embryo development and ageing in birds and mammals', *Proceedings of the Royal Society of London. Series B*, 2006, 273: 2077–82.

Ridgway, R., 'The Birds of North and Middle America', *Bulletin of the United States National Museum*, 1901, 50.

Robson, J., and Lewer, S. H., *Canaries, Hybrids and British Birds in Cage and Aviary*. London: Waverley Books, 1911.

Roger, J., *Buffon*. Ithaca: Cornell University Press, 1997.

Romanes, G. J., *Animal Intelligence*. London: Kegan, 1885.

Romanoff, A. L., *The Avian Embryo*. New York: Macmillan, 1960.

— and Romanoff, A. J., *The Avian Egg*. New York: Wiley, 1949.

Rothschild, W., *Extinct Birds*. London: Hutchinson, 1907.

Runeberg, J., 'The Lark', *Academy*, 1874, 4: 262.

Salvin, O., and Godman, F. D., *Birds of Central America*. London: 1879–1904.

Sauer, F., Stummvoll, J., and Fiedler, R., *De Arte Venandi cum Avibus. Facsimile et Commentarium*. Graz: Akademische Druck, 1969.

Sayle, C., ed., *The Works of Sir Thomas Browne*. Edinburgh: Grant, 1927.

Schaeffer, J. C., *Elementa ornithologia iconibus*. 2nd edn. Ratisbonae: Typis Breitfeldianis, 1779.

Schäfer, E. A., 'On the incidence of daylight as a determining factor in bird migration', *Nature*, 1907, 77: 159–63.

Schierbeek, A., *Opuscula selecta Neerlandicorum de arte medica*. Amsterdam: Van Rossen, 1955.

Schmidtt, S., *Oeuvre*. Paris: Gallimard, 2007.

Schramm, M., 'Frederick II of Hohenstaufen and Arabic science', *Science in Context*, 2001, 14: 289–312.

Schulze-Hagen, K., Leisler, B., Birkhead, T. R., and Dyrcz, A., 'Prolonged copulation, sperm reserves and sperm competition in the aquatic warbler *Acrocephalus paludicola*', *The Ibis*, 1995, 137: 85–91.

Schulze-Hagen, K., Leisler, B., Schaffer, H. M., and Schmidt, V., 'The breeding system of the aquatic warbler *Acrocephalus paludicola* – a review of new results', *Vogelwelt*, 1999, 120: 87–96.

Schut, E., Hemmings, N., and Birkhead, T. R., 'Parthenogenesis in a passerine bird, the zebra finch *Taeniopygia guttata*', *The Ibis*, 2008, 150: 197–9.

Schwenckfeld, C., *Theriotropheum Silesiae*. Lignicii, 1603.

Seibt, U., and Wickler, W., 'Individuality in problem solving: string pulling in two *Carduelis* species (Aves: Passeriformes)', *Ethology*, 2006, 112: 493–502.

Selby, P. J., *Illustrations of British Ornithology*. Edinburgh: W. H. Lizars, 1825–41.

Selous, E., *Bird Watching*. London: Dent, 1901.

— *Realities of Bird Life*. London: Constable, 1927.

— *Evolution of Habit in Birds*. London: Constable, 1933.

Shoberl, F., *The Natural History of Birds*. London: Harris, 1836.

Short, R. V., 'The testis: the witness of the mating system, the site of mutation and the engine of desire', *Acta paediatrica. Supplementum*, 1997, 422: 3–7.

Smellie, W., *The Philosophy of Natural History*, volume 1. London, 1790.

— *The Philosophy of Natural History*, volume 2. London, 1799.

Smith, H. G., and von Schantz, T., 'Extra-pair paternity in the European starling: the effect of polygyny', *Condor*, 1993, 95: 1006–15.

Solinas, F., *L'Uccelliera: Un libro di arte e di scienza nella roma dei primi lincei*, 2 volumes. Florence: Leo S. Olschki Editore, 2000.

Spamer, O., *Illustrirtes konversations-Lexikon III*. Leipzig: 1893.

Stanley, E., *A Familiar History of Birds: Their Nature, Habits and Instincts*, 2 volumes. London: Longmans, 1835.

Stanley, T., *Poems by Thomas Stanley, Esquire*. 1651.

Stap, D., *Birdsong*. Oxford: Oxford University Press, 2005.

Stillingfleet, B., *Miscellaneous Tracts*. London, 1762.

Stresemann, E., 'Baron von Pernau, pioneer student of bird behavior', *The Auk*, 1947, 64: 35–52.

— *Die Entwicklung der Ornithologie. Von Aristoteles bis zur Gegenwart*. Berlin: Peters, 1951.

— *Ornithology from Aristotle to the Present*. Harvard: Harvard University Press, 1975.

Strindberg, A., *En blå bok*. Stockholm: Björck & Börjesson, 1907.

Stubbs, G., *Comparative Anatomy*. London: Orme, 1804–6.

Susemihl, J. C., and Susemihl, E., *Die Vögel Europa's*. Stuttgart: 1839–52.

Suthers, R. A., 'Contributions to birdsong from the left and right sides of the intact syrinx', *Nature*, 1990, 347: 473–7.

Swainson, C., *The Folk Lore and Provincial Names of British Birds*. London: Dialect Society, 1886.

Taber, E., 'Intersexuality in birds.' In Armstrong, C. N., and Marshall, A. J., eds, *Intersexuality*. London: Academic Press, 1964: 287–310.

Tebbich, S., Taborsky, M., Fessl, B., and Blomqvist, D., 'Do woodpecker finches acquire tool-use by social learning?' *Proc. Roy. Soc. London. Series B: Biological Sciences*, 2001, 268: 2189–93.

Tegetmeier, W. B., *Pigeons: Their Structure, Varieties, Habits, and Management*. London: Routledge, 1868.

Teixeira, D. M., 'Plumagens aberrantes em psittacidae neotropicais', *Revista Brasileira de Biologia*, 1985, 45: 143–8.

— 'Perspectivas da etno-ornitologia no Brazil: o exemplo de um estudo sobre a "Tapiragem"', *Boletim do Museu Paraense Emilio Goeldi. Zoologia*, 1992, 8: 113–21.

Temminck, C. J., and Schlegel, H., *Fauna Japonica – Aves*. Lugdun: Batavorum, 1845–50.

Thielcke, G., 'Neue Befunde bestätigen Baron Pernaus (1660–1731) Angaben über iautäusserungen des Buchfinken (*Fringilla coelebs*)', *Journal für Ornithologie*, 1988, 129: 55–70.

Thienemann, F. A. L., *Einhundert Tafeln colorirter Abbildungen von Vogeleiern*. Leipzig: 1845–54.

Thomas, K. *Man and the Natural World*. London: Allen Lane, 1983.

Thompson, A. L., 'Photoperiodism in bird migration', *The Auk*, 1924, 41: 639–41.

Thorpe, W. H., *Bird-Song*. Cambridge: Cambridge University Press, 1961.

— David Lambert Lack. 1910–1973. *Biographical Memoirs of Fellows of the Royal Society*, 1974, 20: 271–93.

— *The Origins and Rise of Ethology*. Heinemann: London, 1979.

Ticehurst, N. F., Letter to the Editors. *British Birds*, 1934, 27: 308.

Tinbergen, N., 'The behavior of the snow bunting in spring', *Transactions of the Linnaean Society of New York*, 1939, 5: 1–95.

Topsell, E., *The Fowles of Heauen or History of Birds*. Austin: University of Texas Press; 1972.

Travies, E., *Les Oiseaux les Plus Remarkables*. Paris: 1857.

Trivers, R. L., *Natural Selection and Social Theory*. Oxford: Oxford University Press, 2002.

Turner, W., *A Short and Succinct Account of the Principle Birds Mentioned by Pliny and Aristotle*. Cologne, 1544.

Vallet, E., Beme, I., and Kreutzer, M., 'Two-note syllables in canary songs elicit high levels of sexual display', *Animal Behaviour*, 1998, 55: 291–7.

Valli da Todi, A., *Il canto de gl'Augelli. Opera nova. Dove si dichiara la natura di sessanta sorte di Uccelli, che cantano per esperienza, e diligenza fatta piu volte. Con il modo di pigliarli con facilita, & allevarli, cibarli, domesticarli, ammaestrarli e guariril delle infermita, che a detti possono succedere. Con le loro figure, o vinti sorte di caccie, cavate dal naturale da Antonio Tempesti*. Rome: N. Mutii, 1601.

Van Drimmelen, G. C., '"Spermnests" in the oviduct of the domestic hen', *J. South African Vet. Med. Assoc.*, 1946, 17: 42–52.

Van Oordt, G. J., and Junge, G. C. A., 'Die hormonale Wirkung der Gonaden auf Sommer- und Prachtkleid. III', *Wilhelm Roux' Arch. Entwicklungsmech. Org.*, 1936, 134: 112–21.

Van Rhijn, J. G., *The Ruff*. London: Poyser, 1991.

Van Zomeren, K., *Klein Kanoetenboekje*. Utrecht: KNNV Uitgeverij, 2003.

Vieillot. L. P., *Histoire naturelle des plus beaux oiseaux chanteurs de la zone torride*. Paris, 1805–9.

Wagner, R., *Fragmente zur Physiologie der Zeugung, vorzüglich zur mikroskopischen Analyse des Sperma*. München: Bayerische Akademie der Wissenschaft, 1836.

— *Icones Zootomicae. Handatlas zur Vergleichenden Anatomie*. Leipzig: Voss, 1841.

Wallace, A. R., review [of *The Intelligence and Perfectibility of Animals from a Philosophic Point of View. With a Few Letters on Man*. by Charles Georges Leroy, 1870], *Nature*, 1871, 3: 182–3.

Wallace, D. I. M., *Beguiled by Birds*. London: Helm, 2004.

Walters, M., *A Concise History of Ornithology*. London: Helm, 2003.

Walton, I., *The Compleat Angler*. London: 1653.

Ward, J., *British Ornithology or Birds of Passage*. Maidstone: Masters, 1871.

Waring, S., *The Minstrelsy of the Woods*. London: Harvey & Darton, 1832.

Weir, A. A. S., Chappell, J., and Kacelnik, A., 'Shaping of hooks in New Caledonian crows', *Science*, 2002, 297: 981.

Welty, J. C., *The Life of Birds*. Philadelphia: W. B. Saunders, 1962.

West, M. J., and King, A. P., 'Mozart's Starling', *American Scientist*, 1990, 78: 106–14.

Westneat, D. F., Frederick, P. C., and Wiley, R. H., 'The use of genetic markers to estimate the frequency of successful alternative reproductive tactics', *Behavioral Ecology and Sociobiology*, 1987, 21: 35–45.

Westneat D. F., Stewart, I. R. K., 'Extra-pair paternity in birds: Causes, correlates, and conflict', *Annual Review of Ecology, Evolution, and Systematics*, 2003, 34: 365–96.

White, G., *The Natural History of Selborne*, 1789.

White, T. H., *The Bestiary: A Book of Beasts*. New York: Putnam, 1954.

Whitteridge, G., *Disputations Touching the Generation of Animals*. Oxford: Blackwell, 1981.

Wickede, F. van, *Kanari-uitspanningen of Nieuwe verhandeling van de kanari-teelt*. Amsterdam: 1786: 96.

Williams, G. C., *Adaptation and Natural Selection*. Princeton: Princeton University Press, 1966.

— 'Natural selection, the cost of reproduction and a refinement of Lack's principle', *American Naturalist*, 1986, 100: 687–90.

Wilson, A., and Bonaparte, C. L., *American Ornithology*. London & Edinburgh: Cassell, Petter & Galpin, 1832.

Wilson, H. R., 'Physiological requirements of the developing embryo: temperature and turning.' In Tullet, S. G., ed., *Avian Incubation*: Poultry Science Symposium 22, 1991.

Witchell, C. A., *The Evolution of Bird-Song, with Observations on the Influence of Heredity and Imitation*. London: A&C Black, 1896.

Witschi, E., 'Seasonal sex characters in birds and their hormonal control', *Wilson Bulletin*, 1935, 47: 177–88.

Wolf, J., *Abbildungen und Beschreibungen merkwuerdiger naturgeschichtlicher Gegenstaende*. Nuremberg: Tyroff, 1818.

Wolff, C. F., *Theoria Generationis*. Halle, 1774.

Wolfson, A., 'Sperm storage at lower-than-body temperature outside the body cavity of some passerine birds', *Science*, 1954, 120, 68–71.

Wood, C. A., and Fyfe, F. M., eds, *The Art of Falconry, being De Arte Venandi cum Avibus of Frederick II of Hohenstaufen*. Stanford: Stanford University Press, 1943.

Wood, N., *British Song Birds: Being popular Descriptions and Anecdotes of the Choristers of the Groves*. London: J. W. Parker, 1836.

Woods, C. P., and Brigham, R. M., 'The avian enigma: "hibernation" by common poorwills (*Phalaenoptilus nuttalli*).' In Barnes, B. M, and Carey, C., eds, *Life in the Cold: Evolution, Mechanisms, Adaptation and Application*, 12th International Hibernation Symposium. Fairbanks: Institute of Arctic Biology, 2004.

Würdinger, I., Jürgen Nicolai, 1925–2006. *The Ibis*, 2007, 149: 198–9.

Wynne-Edwards, V. C. *Animal Dispersion in Relation to Social Behaviour*. Edinburgh: Oliver & Boyd, 1962.

Xamarro, J. B., *Conocimiento de las Diez Aves menores de jaula, su canto, efermedad, cura y cria. Compuesto por Iuan Bautists Xamarrõ, residente en Corte*. Madrid: Imprenta Real, 1604.

Zann, R. A. *The Zebra Finch: A Synthesis of Field and Laboratory Studies*. Oxford: Oxford University Press, 1996.

Zorn, J. H., *Petino-Theologie*, volume 1 (1742) Pappenheim; volume 2 (1743) Schwabach, 1742, 1743.

Glossary

Adaptation A trait (process, behaviour or structure) that confers a selective advantage in a particular environment.

Air sac Part of the bird's respiratory system, not involved in gas exchange. Most birds have nine air sacs.

Autosome A chromosome other than a sex chromosome (q.v.).

Behavioural ecology Study of the behaviour and ecology within an evolutionary framework, explicitly individual selection (q.v.).

Brood parasite Bird (such as the European cuckoo) that parasitises the parental care of another bird species.

Chalazae The two twisted strands of albumen that hold the yolk (ovum) in place inside an egg (singular: chalaza).

Cicatricule An old term for the germinal disc of a freshly laid egg.

Circannual rhythm Cycles of behaviour, regulated by an internal biological clock, that occur on an annual basis.

Classification The assigning of evolutionary relationships and names to birds and other organisms (q.v. taxonomy).

Cloaca The cavity into which the alimentary, urinary and genital ducts open in birds.

Cloacal protuberance The cloacal region of a male passerine bird formed mainly by the seminal glomera (q.v.).

Clock Also referred to as an internal, physiological or biological clock: a biochemical time-keeping mechanism.

Cockatrice A mythical creature, half snake, half cockerel, hatched from eggs laid by a cockerel.

Coitus Another term for copulation.

Colour rings Plastic rings (in US, bands) placed on a bird's legs for individual identification.

Comparative study A comparison (usually statistical) between species and usually controlling for phylogeny, used to infer adaptation.

Conspecific Of the same species (c.f. heterospecific; of another species).

Cooperative breeding A system of breeding in which additional individuals assist a breeding pair in reproduction.

Crystallised song The final, stable stage of song development.

Density dependence Factors whose influence varies according to the density of a population.

Emblem A kind of puzzle with a moral message, generally comprising a title, an illustration (often of an animal) and a brief explanation, usually in verse.

Epididymis The coiled tube in which sperm mature or are stored. In mammals it is attached to the testis; in birds the equivalent structure is referred to as the seminal glomera and is located away from the testes, usually in the cloacal region.

Epigenesis The idea that an organism develops by the gradual differentiation of a fertilised egg (c.f. preformation).

Ethology A term used predominantly from the 1950s to the 1970s for the study of animal behaviour (mainly proximate aspects).

Fertilisation Union of the male and female gametes (q.v. insemination).

Gametes Sex cells: spermatozoa (or sperm) in males; ova (singular: ovum) in females.

Generation An old term for reproduction and embryo development.

Germ cells Cells in an embryo that give rise to gametes (q.v.).

Gonadotrophin A hormone secreted by the pituitary gland that stimulates the gonads and controls reproductive activity.

Group selection The premise that selection operates on entire groups or species rather than individuals.

Half-sider A hermaphrodite in which an individual is male down one side of the body and female on the other. Also known as a lateral gynandromorph (q.v.).

Heterogametic Refers to the sex with two different sex chromosomes; in birds, the female bears ZW sex chromosomes and is the heterogametic sex (in mammals it is the male (XY)).

Higher vocal centre A specific region of the brain responsible for song-learning in passerine birds.

Hypothalamus A gland within the brain that controls digestive and reproductive systems and regulates many basic behaviours, including feeding.

Imprinting A form of learning, usually within a sensitive period (a specific time window) and usually early in an individual's life. Filial imprinting is that in which offspring learn who their parents are; sexual imprinting is that in which individuals learn to identify an appropriate sexual partner.

Infundibulum Region of female reproductive tract (oviduct) nearest the ovary; a funnel-like structure that captures the ovum as it is released from the ovary.

Insemination Transfer of semen from the male to female (c.f. fertilisation).

Instinct A behaviour with a strong genetic component that requires little if any learning (also referred to as innate).

Intersexuality The state of being neither one sex nor the other; includes hermaphrodites.

Lateral gynandromorph A hermaphrodite in which an individual is male down one side of the body and female on the other. Also known as a half-sider (q.v.).

Lek A courtship arena in which male birds gather to display, usually on tiny territories. A lek mating system is one in which males are polygynous and male reproductive success highly skewed towards a few individuals.

Life-history traits Traits associated with birth, death and reproduction, such as clutch size, age of first breeding or longevity that are often linked. Birds such as seabirds typically have a small clutch size and do not start breeding until they are several years old and are long-lived. In contrast, small birds often exhibit the opposite pattern.

Longevity Lifespan; the length of life.

Mating system The number of partners and type of sexual relationship individuals of a particular species typically have.

Migratory restlessness The agitation (wing-whirring, hopping) exhibited by a caged individual during the time it would normally be migrating – also known as *Zugunruhe*.

Monogamy The most common mating system in birds in which one male and one female rear offspring together. Social monogamy refers to the social relationship; sexual (or genetic) monogamy refers to the genetic relationship. A species or a pair may exhibit social monogamy but not sexual monogamy if either engages in extra-pair copulations.

Natural selection A process by which evolution occurs through the differential survival of better-adapted individuals.

Natural theology See physico-theology.

Neuron A nerve cell.

Oviparous Egg-laying (as in birds or many reptiles).

Ovist See preformation.

Ovum The female gamete (plural, ova).

Parthenogenesis A process by which an individual develops from an unfertilised egg.

Passerine Also known as perching birds, or, less precisely, as songbirds. Passerines comprise more than half of all birds (c.f. non-passerines); they include the true songbirds and suboscines, such as the New World flycatchers.

Personality Non-random, individual behavioural specialisations sometimes referred to as 'behavioural syndromes' or 'coping styles' (reflecting how individuals cope with different challenges).

Photoperiod The relative amount of light and dark in a twenty-four-hour period.

Phylogeny The presumed evolutionary relationships between species or other taxa (e.g. genera, families).

Physico-theology Belief that God was responsible for the fit between an individual and the environment.

Plastic song An intermediate stage in song-learning, that occurs after subsong (q.v.).

Preformation The idea common in the seventeenth century that a miniature, preformed individual exists inside either a sperm or an ovum. Those that believed in this were either spermists or ovists, respectively (q.v. epigenesis).

Proximate and ultimate factors Proximate factors are environmental factors that cause effects. Ultimate factors are those responsible for evolutionary effects.

Radioimmunoassay A method that uses the competition between radiolabelled and unlabelled substances in an antigen-antibody (immune-type) reaction to establish the concentration of the unlabelled substance.

Selection thinking Natural or sexual selection operating on individuals rather than groups, populations or species (c.f. group selection).

Semen A mixture of spermatozoa and seminal fluid.

Seminal glomera The paired highly coiled distal ends of the vas deferens that lie inside the cloacal protuberance of male passerines (singular: seminal glomerus). The seminal glomera are the male's sperm store.

Sensitive period A time period when an individual is particularly receptive to particular environmental influences (c.f. imprinting).

Sex chromosomes Distinctive pair of chromosomes that differ in males and females (q.v. autosomes).

Sexual reproduction Reproduction occurring through the fusion of a sperm and an egg (or ovum), allowing for the recombination of genetic material from two individuals, male and female, to create a new and unique individual. The variation between individuals generated by sexual reproduction is the raw material on which natural selection operates.

Sexual selection A process by which evolution occurs through differential reproductive success mediated by male–male competition or female choice. Sexually selected traits are those that enhance an individual's reproductive success through male competition and female choice.

Sonogram An image of sound produced by a machine referred to as sonograph or sound spectrograph, in terms of frequency (or pitch) and duration (along the vertical axis); darkness of the images indicates volume: used to analyse birdsong.

Sperm Male gamete (abbreviation of spermatozoa, singular: spermatozoon).

Sperm competition Competition between the sperm (or, more correctly, the ejaculates) of two or more males for the fertilisation of a female's ova.

Spermist See preformation.

Sperm storage tubule Microscopic tubule, usually hundreds or thousands in number, located at the utero-vaginal junction in the oviduct of birds, in which females store sperm.

Spontaneous generation A phenomenon in which organisms are created from non-living material.

Stopping A bird-keeper's term for stopping the light a captive bird receives in order to alter its annual cycle.

Subsong Earliest stage in song-learning; precedes both plastic song and crystallised song (q.v.).

Survival With reference to a population or species, the proportion of individuals surviving from one time period to another (usually one year).

Syrinx The sound-producing organ unique to birds.

Systematics Study of the evolutionary relationships between organisms (q.v. taxonomy and classification).

Taxonomy The naming and classification of organisms.

Torpor The state of being torpid, involving a reduction in body temperature to conserve energy.

Treddles Term used in seventeenth century for the chalazae (q.v.).

Utero-vaginal junction Region of the female reproductive tract (oviduct) between the uterus and vagina where the sperm storage tubules (q.v.) are located.

Viviparous Term used to describe the situation in which embryo development is internal (c.f. oviparous).

Zugunruhe See migratory restlessness.

Picture Credits

178–9	Private collection
183	Private collection
186	From the Blacker-Wood Collection, Rare Books and Special Collections Division, McGill University, Montreal, Canada
188	Balfour and Newton Libraries, Cambridge
191	Albertina, Vienna
196	Private collection
199	Private collection
201	From the Blacker-Wood Collection, Rare Books and Special Collections Division, McGill University, Montreal, Canada
204	Balfour and Newton Libraries, Cambridge
207	Balfour and Newton Libraries, Cambridge
208	Private collection
211	Private collection
216–17	Balfour and Newton Libraries, Cambridge
219	Balfour and Newton Libraries, Cambridge (*upper*); Private collection (*lower*)
224	Private collection
229	© All Rights Reserved. The British Library Board
233	Balfour and Newton Libraries, Cambridge
238	Private collection
241	Copyright © Patrimonio Nacional
245	Balfour and Newton Libraries, Cambridge
248	Balfour and Newton Libraries, Cambridge
251	Balfour and Newton Libraries, Cambridge
254–5	Albertina, Vienna
261	Private collection
262	© All Rights Reserved. The British Library Board
263	Thorpe (1961)
265	From the Blacker-Wood Collection, Rare Books and Special Collections Division, McGill University, Montreal, Canada
272	Licence granted courtesy of the Rt Hon. The Earl of Derby, 2008
275	Courtesy of the French Ornithological Society
276	With the permission of the Biblioteca Universitaria di Bologna
281	Balfour and Newton Libraries, Cambridge
284	Balfour and Newton Libraries, Cambridge
289	Balfour and Newton Libraries, Cambridge
292	Courtesy of D. Griffin & B. Skinner, with permission
295	Private collection
299	Balfour and Newton Libraries, Cambridge
300	Private collection

Index

NOTE: Page numbers in *italic* refer to illustrations and captions

A NOTE ON THE TYPE

The text of this book is set in Linotype Goudy Old Style. It was designed by Frederic Goudy (1865–1947), an American designer whose types were very popular during his lifetime and particularly fashionable in the 1940s. He was also a craftsman who cut the metal patterns for his type designs, engraved matrices and cast type.

The design for Goudy Old Style is based on Goudy Roman, with which it shares a 'hand-wrought' appearance and asymmetrical serifs, but unlike Goudy Roman its capitals are modelled on Renaissance lettering.